Kaushik Kumar, Divya Zindani

Engineering Materials Characterization

Also of Interest

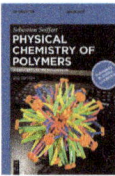

Physical Chemistry of Polymers.
A Conceptual Introduction
Sebastian Seiffert, 2023
ISBN 978-3-11-071327-5, e-ISBN 978-3-11-071326-8

Surface Characterization Techniques.
From Theory to Research
Rawesh Kumar, 2022
ISBN 978-3-11-065599-5, e-ISBN 978-3-11-065648-0

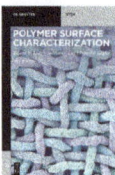

Polymer Surface Characterization
Luigia Sabbatini, Elvira De Giglio, 2022
ISBN 978-3-11-070104-3, e-ISBN 978-3-11-070114-2

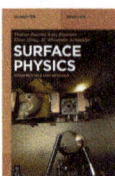

Surface Physics.
Fundamentals and Methods
Thomas Fauster, Lutz Hammer, Klaus Heinz, M. Alexander Schneider, 2020
ISBN 978-3-11-063668-0, e-ISBN 978-3-11-063669-7

Kaushik Kumar, Divya Zindani

Engineering Materials Characterization

—

DE GRUYTER

Authors
Dr. Kaushik Kumar
Department of Mechanical Engineering
Birla Institute of Technology
Mesra, Ranchi 835215
Jharkhand
India
kaushik.bit@gmail.com

Prof. Dr. Divya Zindani
Department of Mechanical Engineering
Sri Sivasubramaniya Nadar College of Engineering
Kalavakkam 603110
Tamil Nadu
India
divyazindani@ssn.edu.in

ISBN 978-3-11-099760-6
e-ISBN (PDF) 978-3-11-099759-0
e-ISBN (EPUB) 978-3-11-098646-4

Library of Congress Control Number: 2023944026

Bibliographic information published by the Deutsche Nationalbibliothek
The Deutsche Nationalbibliothek lists this publication in the Deutsche Nationalbibliografie;
detailed bibliographic data are available on the internet at http://dnb.dnb.de.

© 2024 Walter de Gruyter GmbH, Berlin/Boston
Cover image: Kkolosov/iStock/Getty Images Plus
Typesetting: Integra Software Services Pvt. Ltd.
Printing and binding: CPI books GmbH, Leck

www.degruyter.com

Preface

The authors are pleased to present the book *Engineering Materials Characterization*. The book title was chosen keeping in mind the present trend of material formulation and inventions, and also thinking that a book on material characterization covering various commonly used techniques and instruments would come in handy for various academicians, students, researchers, industrialists, and engineers.

An important aspect of materials science is the characterization of the materials that is produced or developed. The first step in any characterization of a material or an object made of a material is often a macroscopic observation. This is simply looking at the material with the naked eye. This simple process can yield a large amount of information about the material such as the color of the material, its luster (whether it displays a metallic luster), its shape (whether it displays a regular, crystalline form), its composition (whether it is made up of different phases), its structural features (whether it contain porosity), etc. Often, this investigation yields clues as to what other tests could be performed to fully identify the material or to solve a problem that has been experienced in use. However, macroscopic investigations are unable to provide information about defects below the surface like cracks, pores, etc. Hence, microscopic visualization is required. This is also required when the situation is beyond the visual range of naked eye.

Material characterization is a fundamental process in the field of materials science without which no scientific understanding of engineering materials could be ascertained. The scale of the structures observed in materials characterization ranges from angstroms, such as in the imaging of individual atoms and chemical bonds, up to centimeters, such as in the imaging of coarse grain structures in metals.

Microscopy is a technique that, combined with other scientific techniques and chemical processes, allows the determination of both the composition and the structure of a material. It is essentially the process of viewing the structure, already stated, on a much finer scale than is possible with the naked eye. Moreover, it is necessary because many of the properties of materials are dependent on extremely fine features and defects that are only possible to observe using such instruments.

While many characterization techniques have been practiced for centuries, such as basic optical microscopy, new techniques and methodologies are constantly emerging. In particular, the advent of the electron microscope and mass spectrometry in the twentieth century has revolutionized the field, allowing the imaging and analysis of structures and compositions on much smaller scales than was previously possible, leading to a huge increase in the level of understanding as to why different materials show different properties and behaviors.

Spectroscopic instruments are essential tools in material science and research, enabling scientists to study and understand the composition, structure, and properties of various materials. These instruments use the principles of spectroscopy, which in-

https://doi.org/10.1515/9783110997590-202

volves analyzing the interaction between matter and electromagnetic radiation, to gather valuable information about materials at the atomic and molecular levels.

Spectroscopic instruments play a pivotal role in material characterizations, enabling researchers to gain valuable insights into the properties and structures of diverse materials. The information obtained from these instruments aid in various scientific fields, including chemistry, physics, biology, and materials science. As technology advances, spectroscopic techniques continue to evolve, providing even greater capabilities for material analysis and pushing the boundaries of scientific understanding.

The book has two parts, namely, **Part A: Microscopic Instruments** and **Part B: Spectroscopic Instruments.** In both the sections prominent and most commonly used instruments have been discussed.

Part A, i.e., **Microscopic Instruments** presents **five chapters** dealing with **five** most commonly used microscopic instruments for material characterization.

The book commences with **Chapter 1** which introduces readers to the first instrument used after naked eye, i.e., an **optical microscope**. The maximum magnification of such instruments is usually ×1,500, and their maximum resolution is about 200 nm due to the wavelength of light. The advantage of such instrument is that it can be used to view a variety of samples, including whole living organisms or sections of larger plants and animals. This chapter lays emphasis on the contrast mechanism that are particular of microstructural morphologies observed by optical microscopy. Efforts have been made to provide explanations at the elementary levels through the consideration of interaction between the specimen sample and the incident beam of light.

Chapter 2 introduces **transmission electron microscopy**. A **transmission electron microscope (TEM)** utilizes energetic electrons to provide morphologic, compositional, and crystallographic information on samples. At a maximum potential magnification of 1 nanometer, TEMs are the most powerful **microscopes**. The resolutions that are available for morphological studies are extended through the employability of electron microscope. The first attempt to focus the beam of electrons using electrostatic and electromagnetic lenses was accomplished in the 1920s. There have been continuing efforts, which finally gave rise to electron microscope in the 1930s. This was pioneered by Ruska working in Berlin. The electron microscopes were referred to as transmission electron microscope and were employed for thin films, powders, and the sections separated from the bulk material. It was in the Second World War that the reflection electron microscope was pioneered that were capable of producing images for solid samples at glancing incidence. This chapter outlines the associated principles that are employed in focusing an image with high energy beams of electrons. The requirements for transmission microscopy and the detailed specimen preparation procedure are detailed next.

Chapter 3 goes a bit deeper and takes up **scanning electron microscope**. It is a type of electron microscope that produces images of a sample by scanning the surface with a focused beam of electrons. The electrons interact with atoms in the sample, producing various signals that contain information about the surface topography and composition of the sample. Scanning electron microscope has gained popularity owing to its potential ability to reveal the detailed information of those that are displaced along the optic axis in addition to the two-dimensional view of the image produced in the plane of the image. The chapter also reveals the information on the interpretation of the information in the signals in tandem to the special available techniques to enhance the contrast of the image and assist in interpretation of the image.

The next chapter, i.e., **Chapter 4** goes still deeper into the morphology of the material and introduces **scanning probe microscope**. It is a technique and the results that arises when a sharp, needle-shaped solid probe replaces the electromagnetic radiations or the high-energy electron beam. In this chapter, it will be learnt that the employability of such a probe can produce additional information associated with the microstructure of the specimen under investigation. The associated techniques reveal the properties associated with both the structure and the properties of the structure. The information available from scanning probe microscopy have proved to be advantageous for the range of engineering problems in the domain of adhesion and lubrication, biological domain, electro-optical devices, and the investigations associated with the sub-micrometer electronics.

The last chapter of Part A, **Chapter 5**, talks about **X-ray diffraction analysis**. X-ray diffraction, frequently abbreviated as **XRD**, is used to analyze the structure of crystalline materials. Hence, it is used to identify the crystalline phases present in a material and thereby revealing chemical composition information. Confidence with which the chosen model can be considered to be equivalent to the crystal structure will depend on the extent to which the predicted spectrum fits the measure data.

The book then starts with **Part B**, i.e., **Spectroscopic Instruments**. The first chapter of this part, **Chapter 6**, introduces readers to **Fourier transform infrared (FT-IR) spectrometer**. FT-IR spectrometer analyzes the interactions between infrared light and a material. They provide information about molecular vibrations and functional groups present in a sample. FT-IR offers various advantages based on the material being analyzed. The salient feature being identification of organic and inorganic compounds, detection of impurities in materials, analysis of polymers and biomolecules, etc. to name a few.

The next chapter, i.e., **Chapter 7** discusses another very common yet very important instrument named **Raman spectrometer** which bears the name of the inventor **Dr. C. V. Raman**. Raman spectroscopy involves the scattering of monochromatic light by a sample, resulting in a shift in energy that provides insights into molecular vibrations. Raman spectrometers are powerful tools for analyzing crystalline materials and identifying molecular structures, even in complex matrices. The unique characteristics being characterization of crystal structures and lattice vibrations, identification of minerals and gemstones, analysis of carbon-based materials such as graphene, and so on.

Chapter 8 provides information about **X-ray photoelectron spectroscopy**. X-ray photoelectron spectroscopy (XPS), also known as electron spectroscopy for chemical analysis (ESCA), is a powerful surface-sensitive spectroscopic technique used to analyze the elemental composition and chemical states of materials. XPS provides valuable information about the outermost atomic layers (typically 1–10 nanometer) of a sample, making it particularly useful for studying surfaces, thin films, and interfaces.

In **Chapter 9, ultraviolet photoelectron spectroscopy (UPS)** has been discussed elaborately. UPS is a surface-sensitive spectroscopic technique that provides valuable information about the electronic structure of materials, particularly the energy levels of valence electrons. UPS is based on the photoelectric effect, similar to X-ray photoelectron spectroscopy (XPS), but it uses ultraviolet (UV) photons instead of X-ray photons to excite electrons.

The penultimate chapter of the book and the part, **Chapter 10**, elaborately explains **fluorescence spectroscopy**. Fluorescence spectroscopy is a versatile and widely used technique in scientific research, providing valuable information about the structure, dynamics, and interactions of molecules. It is used to study the fluorescent properties of molecules. Fluorescence spectroscopy involves the absorption of light at a specific wavelength and the subsequent emission of light at a longer wavelength, a process known as fluorescence. Its applications span across biology, chemistry, environmental science, and materials research, making it an indispensable tool for modern analytical and investigative work.

The last chapter of Part B and the book, **Chapter 11**, describes **nuclear magnetic resonance spectroscopy**, more commonly known as **NMR**. NMR spectroscopy is based on the principles of nuclear magnetic resonance and provides detailed information about the local environment of specific nuclei within a material. It is particularly useful for elucidating the structure and dynamics of organic and inorganic compounds in solution or the solid state. The technique can be used for determination of molecular structures and conformational changes, study of protein folding and interactions, analysis of complex mixtures such as petroleum products, and many other applications.

God, with your kind blessings this work could be completed to our satisfaction. Your gift of writing and power to believe in passion, hard work, and pursue dreams had made it possible. This could never have been done without the faith in You, the Almighty. Thank you for everything.

We would like to thank our grandparents, parents, and relatives for allowing us to follow our ambitions. Our families showed patience and tolerated us for taking yet another challenge which decreases the amount of time we could spend with them. They are our inspiration and motivation. Our efforts will come to a level of satisfaction if we can make researchers and students with nonmaterial science background understand and work on these characterization instruments and get benefitted.

The authors would also like to thank the Reviewers, the Editorial Board Members, Project Development Editor, and the complete team of De Gruyter for their constant and consistent guidance, support, and cooperation at all stages of this project. Their effort in every phase, from inception to execution, cannot be expressed in words.

Throughout the process of writing this book, many individuals, from different walks of life, have taken time out to help. Last, but definitely not the least, the authors would like to thank them all for providing them encouragement. The project would have got shelved without their support.

Kaushik Kumar
Divya Zindani

Contents

Part A: **Microscopic instruments**

Chapter 1
Optical microscopy

1.1 Introduction

Morphological characterization forms one of the important elements in the domains of science, engineering, and medicine and optical microscope is the primary tool to accomplish it. Thin slices of biological tissues are prepared for transmission electron microscopy and the microscopy is performed with additional image contrast through the aid of darkfield optical microscopy, fluorescent dyes, phase contrast or differential interference. Geology is another domain that works on transmission microscopy. The mineralogical specimens are cut down to thicknesses less than 50 μm and then the polished specimens are mounted on the transparent glass slides. In this case, the major source of contrast is the anisotropy of the sample that is viewed under the influence of polarized light. As such the information is also provided on the spatial orientation and optical properties of the present crystalline phase in addition to the morphological characteristics.

Henry Clifton Sorby is credited for preparing the metallurgical samples in the form of thin slices for carrying out metallographic examination [1]. A similar methodology was used to carry out the examination as that employed for the mineralogical samples. However, all the metallurgical samples are required to be examined in reflection owing to the opaqueness rendered to the metals by the conducting electrons. As such only the images encompassed the information associated with the surface of the samples and hence it was concluded that the contrast seen in the images could be attributed to the optical properties and topology of the specimen surface. The contrast in images in the case of reflection microscopy could be because of surface topology, or due to the difference in assimilation of the light being incident onto the surface, or may be because of the result produced because of optical effects produced due to optical interference and reflection.

In the case of polymers, the images can be obtained either by reflection or transmission. However, the glassy or amorphous phase leads to poor quality images. On the other hand the crystalline phases in the polymers are studied in transmission and are accomplished by casting slim coatings of polymers in melted state on the glass-slide. The contrast in the images is produced by the polymeric crystals in polarized light and is characteristic of optically anisotropic crystal lattice of polymers. The polymeric composites filled with reinforcements can also be examined through reflection; however, the preparation of the composite specimen is cumbersome because of the differences in mechanical properties of the high-modulus reinforcing agent and low elastic modulus possessing polymeric matrix. The advent and presence of the scanning electron microscope (SEM) has reduced the interest in carrying out examination of such polymeric composites through the optical microscope. On the one hand, the

https://doi.org/10.1515/9783110997590-001

contrast produced in case of scanning electron microscopy is unresponsive to the anisotropy of the material, whereas on the other hand, this forms the principal source of contrast in polarized light microscopy.

Examination through reflection microscopy is done to analyze ceramics and semiconductors. Poor optical contrast is often observed in many cases of ceramic samples when they are viewed by the reflecting light. This is due to the poor reflectivity and the strong absorption of the light incident onto the ceramic samples. Furthermore, ceramics being resistant to chemical etchants, do not allow the microstructure to be revealed for a polished ceramic sample. There occurs an altered reply of the specimen to the sample preparedness in the cases where small quantities of dopants or impurities are present. This is due to the sturdy seclusion of the dopant to the grain boundary as well as interfaces.

The present section lays emphasis on the mechanism that is particular to microstructural morphologies detected by optical microscopy. Efforts have been made to provide explanations at the elementary levels through the consideration of interface between the specimen sample and the incident beam of light. Much of the prominence has been given to the reflection microscope. A majority of the present discussion is also valid for the transmission samples.

1.1.1 Geometrical optics

There has been lesser attraction towards the aspects of geometrical optics owing to the rapid developments that have taken place in the domain of physical science over the past century. The topic has become mundane also because of the technological revolution associated with communications, microelectronics, and semiconductors. The present discussion, however, provides a rudimentary knowledge as to how the optical microscope works.

1.1.2 Formation of optical image

The higher refractive index of glass in comparison to the surrounding atmosphere is responsible for the formation of image by the lens of the optical magnifying glass. Refractive index is determined by the proportion of the incident angle to the transmission angle of the light incident onto the polished glass block. The wavelength of the incident light also reduces as it passes through the glass block.

The convex glass lens leads to deviation of a parallel beam of light that varies with the spacing of the beam from lens axis. The corresponding beam of light is taken to a focus by the convex lens and is termed as the focal length. This length is characteristic of the wavelength of the light. The concave lens on the other hand has negative curvature and causes the incident parallel beam of light to diverge. In this case, the deflected light beam will seem to be created from a point forward-facing the concave lens and the length of the point from the lens is known as negative focal length, $-f$.

The lenses present in optical microscope are combinations of convex as well as concave, possessing different refractive indices aimed at enhancing the presentation of the optical microscope. The position of these lenses inside the optical microscope defines the terms by which they are referred to: objective lens, midway lens, and eyepiece. Also, it is fairly known that the medium surrounding the nearby objective lens and the specimen is liquid. Therefore, an immersion lens is used between the specimen and the objective lens that is intended to be employed with a top refractive index liquid.

The focal distance is constant only for the unicolor light and is dependent on the refractive index of the glass, which varies with wavelength of light. Visible light is dispersed and is not brought to a single focus; whereas blue lights possessing shorter wavelengths are taken to focus on a bigger span in comparison to the red lights possessing higher wavelengths. Therefore, a parallel ray of white light is not focused sharply by the glass lens but to a region referred to as *disc of least confusion*. The phenomenon is known as chromatic aberration. Furthermore, even the monochromatic light may not be focused in case of lenses that are thick and have larger diameter; in such cases a shorter focal length is possessed by the outermost lens regions in comparison to the ones that are closer to the lens axis. As such, a disc of least confusion is again formed and the phenomenon that occurs is referred to as spherical aberration. The aforementioned aberrations can be corrected only with the aid of objective lenses because these lenses can cover for the thickness associated with the glass cover slip. This cover slip protects the samples from the surrounding environment [2].

The optical performance of objective lens is directly related to its cost. The performance has been described by using different technical terms. Achromats are one of the objective lenses that are used to correct the chromatic aberrations at two wavelengths: red and blue. Achromats also have the potential to correct spherical aberration for lights with midway wavelengths such as the green light. The best performance is achieved with monochromatic green light. Plan achromats ensure that the periphery of the lens is also in the view of focus in addition to ensuring the focus for central field of view. These lenses are used to record images instead of just viewing the samples. Plan achromats are the most expensive of all the objective lenses as they are entirely modified for the complete visible range of wavelengths; in addition they also provide in-focus image over the full field of view. Specifically, objective lenses are obtainable that can aid in viewing over a wide range. The prominent examples include: differential interference, darkfield microscopy, phase contrast, etc.

As and when the human eye perceives the visible world, an image of it is created on the retina. The light that travels through the eye's lens is focused to achieve this. By using a network of muscles in the eye to change the curvature of the lens, the eye is able to ensure that objects at various ranges can be brought to sharp focus. However, the focus is age-dependent and deteriorates on aging. The control of muscles is perfect in early childhood. The focus of the human eye can be corrected through prescription spectacles.

The iris controls the light that enters the human eye. The iris works as a flexible aperture that effectively reduces the diameter of bright light lens. The light-sensitive retina is covered in a dense network of optical sensors. Cones and rods function well as optical receptors, detecting incident light with exceptional sensitivity. The retina not just reacts to incident light's intensity, but also to its wavelength, which enables color vision. The response of the human eye to the image formation diminishes as the intensity of the light is reduced and in almost 20% of the humans there is lack of perfect vision. As such the world becomes gray at dusk.

A ray diagram can be used to analyze the image formation in a slim lens approximation. The lens will deflect a parallel beam of light coming from a location in the object plane that is some distance $-u$ in front of it. The focal point f, which is located behind the lens, will let in the light that was refracted. A beam of light coming from the same spot but passing via the conjugate focal point $-f$ in front of the lens, on the other hand, will be bent parallel to the axis of after passing through the lens. If the ray of light passes through the lens, then after passing the lens it will remain undeflected. In all the three cases, the rays of light will meet at a point situated at distance v behind the lens. The magnification of the image produced is given by $M = v/u$ and on the other hand the relationship between the object and image distances can be established as $1/u + 1/v = 1/f$ [2].

1.1.3 Resolution in case of optical microscope

There is change in wavelength of the electromagnetic radiation that passes from the sun to the Earth's atmosphere. The wavelength changes from infrared to ultraviolet. However, the peak intensity of the solar radiations reaching the Earth's surface is in the green region of the light spectrum, that is, visible light. This intensity is too close to the peak intensity of eye. The halide spectrum has been modified through "daylight" filters to simulate sunlight. Particularly, "green filters" are used for observing unicolor radiations and hence gray scale recording [2].

The spatial dispersal of the intensity emanating from the source of light, more particularly, the point of light source that is located at infinity distance, determines the resolution of the optical lens. The calculations relating to the intensity of light assumes that a parallel ray of light travels along the lens axis and is brought to meet at a focal point. For the cylindrically symmetric case, the ratio of intensities in the primary and secondary peaks is 9:1 in the spatial distribution of image intensity. Abbe-equation provides the size of primary peak and is given as follows:

$$\delta = 0.61 \frac{\lambda}{\mu \sin \alpha} \tag{1.1}$$

where λ is the wavelength of the emitted radiation, the lens aperture is defined via α that is obtained by the proportion of lens radius to the focal-length of the lens, and μ is the refractive index of the medium.

1.1.4 Diffraction grating

A diffraction grating comprises of a collection of tightly spaced parallel lines. A cylindrical wave front is generated when the diffraction grating is illumined by a normal and parallel ray of light. This occurs because of the scattering that takes place when the incident light strikes the closely spaced parallel lines in the diffraction grating. The wavelengths generated interfere to produce zero-order and diffracted transmitted beams. The angle of diffraction is given by eq. (1.2) in case the spacing of lines in the grating is too large in comparison to the wavelength of incident light.

$$\sin \theta = n\lambda/d \tag{1.2}$$

A lens can be employed to image-copy a diffraction grating only if the following condition depicted in eq. (1.3) holds:

$$\sin \alpha \geq \sin \theta = \lambda/d \tag{1.3}$$

The above implies that the angular aperture of lens is big and that it can admit both the zero-order as well as the first-order beams. This condition can be compared with the aid of either the Bragg equation for diffraction or with the Abbe relationship. Therefore, all the aforementioned three equations describe the limiting situations for transmission of information about an entity by employing of electromagnetic radiations. The main constraint in all the aforementioned cases is the proportion λ/d. Furthermore, the limit to the resolution limit associated with the microstructural information is directly proportional to the wavelength of the incident electromagnetic radiations [2].

1.1.5 Resolution and numerical aperture

Optical resolution has been defined by Rayleigh as the capability of the lens to differentiate between two point sources at infinity, given that the point sources are viewed in the image planes. The Rayleigh criteria for the microstructural determination is that the angular departure of two light sources of equivalent intensities would confirm that the highest of primary image peak should fall on the least peak from the two different sources of lights respectively. The conditions from Rayleigh suggest that the collective image of the two sources will delineate a small but noticeable intensity that is smallest at the center. It can be concluded here that the Rayleigh criterion agrees on the size of main intensity peak and is similar to that given by the Abbe equation as depicted in eq. (1.1). The Rayleigh resolution is well-defined for the focal plane of the lens system for an image of infinity-point source. The objective lens in the optical microscope magnifies an object, that is, the image is placed far from the lens while the object is in front of the lens. The optical microscope therefore obeys the Abbe equation in that least departure can be differentiated within the object focal plane.

Here, the fundamental significance of the Abbe equation is recalled: no detail associated with the image can be transmitted in case the wavelength being employed to transmit the information is low. This applies to the case wherein the image system is based only on the wave optics. The information associated with the image can be maximized only if the maximum possible optical signal generated from the object can be collected. This signifies that the aperture of objective lens is maximized.

In actuality, a resolution that falls below the wave optical limit is achievable. By using the phonon particle characteristics, this can be accomplished. This is the fundamental building block of near-field microscopy. In this instance, the specimen's surface is scanned with a sub-micrometer light pipe. The resolution is dependent on the light pipe's diameter and distance from the surface. The settlement that is reached may be far superior to the one established by the Abbe relation. The context itself, though, is outside the purview of the debate at hand. Recent developments have helped surpass the resolution limit imposed by diffraction. The fluorescence emission from the histology specimen stained using the fluorescent dye has been slightly quenched to achieve this. Fluorescent emission is stimulated by pulsed laser light. A momentary picosecond laser pulse, which is timed right after the first one, partially quenches the peripheral fluorescent signal. This makes sure that just the light produced from the center is captured. The resolution has been close to 30–40 nm. The mentioned experimental method should only be viewed as proof that it is feasible to surpass the diffraction limit.

As the effectiveness of an optical microscope is not limited, aberrations in the objective lens can be addressed. One of the key features that distinguish the objective lens system is the parameter $\mu\sin\alpha$ often known as the lens's numerical aperture. The highest numerical aperture of the immersed lens system is 1.3, compared to 0.95 for the air-operated lens. On the outer edge of the objective lens numerical aperture numbers are marked.

The resolution limit of objective lens differs from the detection limit and the differences must be discussed. The signal will be difficult to detect if its intensity decreases. In case of the reflection microscope, the light is scattered outside the aperture of the lens for the smallest objects. This results in deficit in the light intensity in and around the field of image view. The object's apparent size, or Abbe width, stays constant while the strength of the signal emanating from it decreases as the object's size gets smaller and smaller. The object will stop being identified at a certain size, or the limiting size, since the signal originated at that size will exceed the detection system's background noise limit. The wavelength of the light and the lens's numerical aperture determine the detection limit, which is less than the resolution limit. One of the major reasons behind the usage of dark field illumination is the easier detection of smaller features that scatter the originating light into the field of image generation. In this case, the reliability on small amount of light scattered is pretty low.

1.1.6 Depth of focus and depth of field

Any object resolution whose image copy is in focus in the image level is limited. The finite resolution is limited by the numerical aperture possessed by the objective lens. Therefore, it can be concluded that objectivity is not required to be at precise object space from lens u, which can be dislocated from the image plane deprived of even losing the resolution. Depth of field defines the space over the object focal stays. This depth of field is given by the following eq. (1.4):

$$d = \delta \tan \alpha \qquad (1.4)$$

In the above equation, α is half the angle subtended via the objectivity of aperture on the point of focus. Similarly, the image of the object will stay in focus if it is dislocated at a certain space v from lens. The depth of focus determines the distance up to which the image stays in focus and is given by eq. (1.5):

$$D = M^2 d \qquad (1.5)$$

where magnification is given by M. Both the aforementioned expressions are based on the assumption that the objective can be considered as "thin lens." This is, however, not the same with the microscope on the market. From the expression $\delta = 0.61\lambda/\mu \sin \alpha = 0.61\lambda/NA$ it is obvious that field depth declines with the increase in NA. It is recommended that for better resolution, the image must be located to an accuracy more than 0.5 μm. This is also one of the essential requirements when the mechanical stability is specified at the sample stage.

The focal depth is considered to be of lesser importance. The movements of millimeter range in the image plane are acceptable given the fact that the amplification larger than × 100 will be required if all the resolved detail is recorded.

1.2 Construction of the microscope

The optical microscope is an assemblage of three distinct arrangements: illuminating system, the specimen stage, and the imaging system. The source of light used for illuminating the samples is provided by the illuminating system. The sample stage holds the sample as well as controls the position of the sample in accordance with area to be observed. The imaging system provides a means to transfer the amplified and undistorted image to the observation plane as well as to the recording medium [2].

For light source, there are two requirements that are conflicting in nature. On the one hand, sufficient light must flood the sample area that is under inspection beneath the objective lens. This will ensure that almost all the microstructural features are subjected to similar illuminating situations. Additionally, incident light has to focus onto the sample in such a manner that the reflected intensity allows for comfortable viewing.

The light source would be as glowing as conceivable. About half a century ago, the striking of carbon arc was revealed to yield excellent results; however, the light source was unstable. On the other hand, monochromatic emission line in green was generated by the employability of mercury arc lamp, the wavelength of which was compatible with the peak sensitivity of a person's eye. In the present scenario, a conventional light bulb is used in small instruments while a tungsten-halide discharge tube is employed in the modern and high-performing optical microscopes. Their employability has provided for a steady and strong source of white light that corresponds to a high temperature of 3,200 K. A narrow band of wavelength can be used by the filters that are usually green or to create an effect of sunlight from close quarters.

Apart from light source, there are additional vital mechanisms in the illuminating system. The image of a source is focused by the condenser lens that is near the back focus of the objective lens plane. This ensures that uniform illumination of the surface takes place by the employing a near parallel light ray. The amount of light that is emitted from the source and is allowed to enter the microscope is limited b effectively. Condenser aperture of small size can be maintained to enhance the contrast in the image; however, this is accomplished at the cost of reduction in the intensity of the image. There is another aperture, which is known as the simulated image aperture, which is positioned in the simulated image plane. This ensures that the light that illuminates the zone underneath inspection is only acknowledged by the microscope. Unwanted background intensity is also discarded by ensuring that the light is not reflected internally. The extent of the simulated image aperture is tuned in accordance with the magnification of the microscope. The condenser and the simulated image apertures are endlessly flexible irises that can be tuned in accordance with the essential sizes [2].

Illuminating systems are repositioned in case of many reflection microscopes. This ensures that the optically see-through samples can be viewed in transmission. This is particularly helpful in the case of slim tissue sample of biology and also used for mineralogical samples, polymers that are partly crystal-like, and thin film samples that are equipped for transmission electron microscopy.The simulated image aperture may be provided alternatively with the central stop, allowing a ring of light to illumine the zone that is to be observed. This darkfield illumination will greatly affect the contrast.

Mechanical stability is one of the essential requirements for the primary stage and therefore the specimen must be positioned accurately in order to better the resolution limit. The accurate positioning of the specimen is just one of the aspects of the mechanical stability. The vertical location of the specimen is adjusted to bring the image into the focus and the accuracy with which the z-adjustment is accomplished should be ensured within arena depth for the major numerical aperture objective lens.

The required mechanical accuracy is achieved usually by the fine in addition coarse micrometer screws for all the three coordinates and it is then required to minimize the time reliant on drift as well as the slack in the system.

It is therefore the z-adjustments that bring forth a number of problems. This is because a rigid stage suggests a fairly enormous and hefty construction. Reliant on that the sample is required to be located above or beneath the objective lens, there are two main possibilities that exist. In the former, the plane of the surface should be located precisely normal to the axis of the microscope. This is accomplished by supporting the sample from beneath with Plasticine and using a suitable jig to apply light pressure.

The type of objective lens accessible depends on the properties of the material as well as the imaging method. The objective lens's performance is defined by its numerical aperture. This typically appears on the side of the objective lens assembly. This is also the magnification value for that lens. As most objective lenses are achromatic, they are not confined to monochromatic light. These are, nevertheless, suggested for seeing high quality monochromatic pictures in green. Achromatic lenses provide a focused picture only in the center of the field of view. Such lenses have the advantage of correcting both spherical aberration in the green and chromatic aberration in the red and blue. The use of plan achromat objective lenses ensures that the edges of the field of view are in sharp focus. The plan achromats are hence most suited for recording images. On the other side, apochromatic objective lenses do not have three-wavelength chromatic aberration [2].

Histologic investigation of soft tissues is one of the main domains of the work in life sciences that is carried out by using the optical microscope. This requires the specimen to be covered from the outside environment and is accomplished by mounting the slim tissue slide of a glass-slide, and then a coverslip is used to protect it. A similar technique is used for many of the polymeric specimens. The materials are cast onto the slide while the spinning controls the thinness of the specimen. This is also accomplished by exerting even pressure to the cover slip. Objective lens systems that are used with such samples are required to be modified for the refractive index. Also, it is required to correct the thickness of the optically flat cover plate.

The numeric aperture of the objective lens determines resolution as well as the brightness of the objective lens. The luster of the image is inversely related to the square of the amplification. The cone acceptance angle, however, increases with the numerical aperture of the objective lens. Owing to the large cone acceptance angle, more light is collected. The numerical aperture of the lens can change by any magnitude order when progressing from low amplification lens to the higher-order magnification system. Then all the associate image details can be viewed by increasing the magnification in the similar order.

As the numeric aperture of the objective lens is increased, the employed space of the lens from outward the specimen declines intensely. For the case of high-powered lenses this goes down to the order of 0.1 mm. The objective lens system could be damaged if the specimen is driven through the focus into the glass lens and replacing lens system is an expensive affair.

Special lenses that cater to long working distances are available. Such lenses allow high-magnification observations with the elimination of the sample brought in too close to the objective lens. A midway image is created at unit amplification by an inexpensive design and even allows the sample to be imaged under unfriendly situations such as the corrosion medium or under cryogenic conditions. A number of cases have attracted in situ experiments; however, only a small number under such dynamic conditions have been carried out using optical microscopy. There are numerous associated difficulties: dimensional stability of the support structure and also the specimen, obstruction of the path between the objective lens assembly, and the specimen by a chemical, etc.

A wide variety of objective lenses and stage assemblies for optical microscopy is available. Darkfield objective is mostly valuable for carrying out both reflection and transmission works. The aforementioned type of objective lens illumines the sample with the funnel of light that surrounds the aperture of lens. A darkfield image is then shaped by the scattering of the light that is taken by the specimen and into the aperture of the lens. The intensity in the case of the darkfield objective is in contrast to that observed in the case of normal illumination.

1.3 Image observation and sensitivity of recording

The magnified image provided by the objective lens assemblage is inadequate to be completely resolved by a person's eye. Therefore, here are three different choices that are accessible and can aid in sufficiently resolving the image provided via the objective lens. One is to insert an eyepiece and an added tube or midway lens. This makes it comfortable for the observer to sight the image at the employed amplification directly. Most of the microscopes are now designed with the duct lens that allows the samples to be located directly into the focus plane of the objective lens. This ensures that the light returned from the objective lens is corresponding and is taken to the focal point by using the tube lens. This also permits for a number and varying range of optical accessories among the objective as well as the duct lenses. In another choice, additional images may be incorporated to focus the image conveniently onto a photographic blend, charge-coupled device that is light-sensitive and supports in subsequent enlargement of the image. In the third option, a television raster can be used to scan the image and display it on the monitor. This option may be the preferred technology for recording the dynamic events into the microscope. Recent advancements have led to the growth of high-value charge-coupled device cameras and has made possible the elimination of objective lens in the process to record the images. High-quality charge-coupled devices have now been incorporated into computerized cameras and have replaced photographic recording and hence the television camera recording technology. Photographic emulsions were, however, used by professional photographers as such cases require the highest possible performing charge-coupled devices that are very costly. More recently, advancements have led to the development of complementary metal oxide semiconductors. This development has resulted in a single chip camera. Individual pixels can be filtered in both the charge-coupled devices as well as complementary metal oxide semiconductors. Both the charge-coupled devices as well as the complementary metal oxide semiconductors offer pixel sizes lesser than 6 μm. However, complementary metal oxide semiconductors are inadequate to the order of million-pixels per frame. The order is less compares to the charge-coupled devices. Plan apochromatic objective lenses is one of the popular domains for investment that allows for high-resolution computerized color recording [2].

Monocular eyepieces can be used for visual observation. The primary image is enlarged by a feature of 3–15. A common non-immersion objective with numerical aperture of around 0.95 has a main amplification of 40 and the resolution is as close to 0.4 μm and ensures that all the resolved features are gladly observable to the eye (0.2 mm) with the requirement of some further magnification that can be calculated as: $200/(0.4 \times 40) = 12.5$.

Further, most decent microscopes incorporate an added midway or duct lens and make it convenient to resolve $\times 3$ and $\times 5$ eyepiece and hence resolve all the image details. Also, care needs to be taken to avoid the usage of very high eyepiece (×15) that would otherwise reduce the arena of sight and expand the resolve feature to a point where they seem blurry to the observer's eye.

A beam splitter or a binocular viewer may also be incorporated by some microscopes and is particularly convenient for the ones that struggle to view through one eye. However, some disadvantages must be known to the microscopist. It is unusual to have identical focal planes in both the eyes and therefore one eyepiece of the couple is required to be focused autonomously. The employer uses one eye to first focus a feature of curiosity in the plane of the sample. The flexible focus of the second eyepiece is then adjusted till the images that are realized by both the eyes merges into a solo and simultaneously focused image. The process of regulating the binocular setting is then finished by making sure that the parting of the two eyepieces tie with the parting of the observer's eye. A binocular eyepiece, however, does not provide for stereoscopic viewing of the sample that ensures three-dimensional observing of the sample. This also requires two objective lenses to focus on the same field of view. Although stereo binoculars that possess twin objective lenses exist, their magnification is restricted to about 50 times. The geometrical issues related to the placement of the objective lenses set the limit. Electronic industries widely employ stereo binoculars that are therefore one of the major tools for viewing the specimen.

The way an image emulsion and a person's retina respond to light differ significantly. Maximum sensitivity is exhibited in the ultraviolet by the emulsions. Also the maximum sensitivity is for black and white. The color pictures rely greatly on the colorants to spread the photosensitivity of silver halide blends outside the green. Orthochromatic emulsions do not showcase sensitivity to red light, which is convenient particularly in darkroom process. These are commonly preferred in case of photographic footage of monochromatic microscopic images in green light. One of the common choices for black and white photographic footage in daytime is that of panchromatic film. However, the sensitivity in this case falls steadily with the increased wavelength. The recoding medium in the case of classical black and white photography yields a negative wherein the black zones correspond to supreme light excitation and the clear areas correspond to zero excitation. Polaroid cameras produce positive grayscale image. Color footage films might be negative which are employed to accomplish color recordings. It is customary to use the suffix "chromite" for good transparencies and the suffix "color" for negative films [2].

The response of emulsion to a fixed dose of radiation at a standard wavelength is referred to as the speed of an emulsion. The speed of emulsion rests on three important factors: the exposure time, the development process, and the "grain" of the emulsion. A photosensitive silver halide grain reacts to successive growth only if there are chances where it can absorb a pair of photons. Since the grain might decline from its early excited state, the time interval between the arrivals of photons is very important. The interval between photon excitations increases with the increase in the incident intensity. The response of the emulsion, on the other hand, reduces. The phenomenon is termed as reciprocity failure. Larger sizes of halide grains enhance the cross-section of the photon collision and improve the photosensitivity. Grainier image with poor inherent resolution is the result of such mechanism. The mechanisms suggest that negotiation is obligatory among slow-speed, fine grained blends and high-speed, coarse-grained blends [2].

The grain of silver during the development phase rises into a group of silver grains and the volume is much larger in comparison to the volume associated with a single activated crystal structure. Thus, the resolution of the last-noted image is greatly affected by the silver grain growth during its developmental phase. An emulsion that is to be employed for photomicroscopy and is designed as per the recommendations should possess a resolution range of 10–20 μm. Moreover, the prepared emulsion must be capable of producing amplification by an aspect of × 10. Hence, it should be feasible to capture high-resolution images without any loss of information at magnifications lower than those needed to observe a fully resolved microstructure.

Thus, the low magnification and high-resolution noted images will contain additional info in comparison to that available in the field of view.

The dose requirement of the blackening determines the contrast achievable in a given blend. The amount of light that is available per unit area (E) when increased by the time of exposure (t) gives the dose. On the other hand, blackening (D) can be defined using the following eq. (1.6):

$$D = \log\left(\frac{I_0}{I}\right) \tag{1.6}$$

where I_0 is the intensity of the light incidence on the blend and I is the strength of the light that is conveyed through the unprotected and advanced blend.

Contrast refers to the slope of the curve obtained by plotting D against log of product of the quantity of light available per unit area and the exposure time with the emulsion. Emulsions with a higher value of contrast lose the detail associated with the image, owing to the tendency to record as white or black with the midway presence of gray levels. On the other hand, emulsions with low contrast lack contrast because of the fact that the gray levels are very close to one another. Another demerit related to photographic recording is the nonlinearity of the response of the emulsion. Another disadvantage is that associated with the difficulty in controlling the number of parameters that are involved in exposing, enlarging, printing, and developing the emulsion. Making quantitative measurements of image intensity is a very difficult task. It is preferable to employ charge-coupled devices to accomplish computerized recording.

There are, however, limitations to recording the range of information either by photographic emulsion or by a charge-coupled device. The background noise becomes a problem at a very low dose. On the other hand, the recording media will start to saturate at very high doses. Four orders of scale of quantity can be responded to by the high resolve, black, as well as white negative picture. On the other hand, the positive print is restricted to only two orders of magnitudes. The charge-coupled devices camera has the distinct advantage that they provide for linear response. The response to high-energy electrons of the photographic emulsions is also linear owing to the fact that the halide grains are excited by the impact of the solo electron in comparison to the excitation procedure of visible light that needs two photons.

Optical microscopes have been attached with television cameras and monitors for long. This arrangement allows viewing the observations made by microscope in real time. It has also aided in recording of dynamic events such as corrosion studies, effect of heating the sample, etc. that happen under the optical microscope. However, the numbers of pixels that are scanned by the television are lower in comparison to the number of image elements that can be resolved by a normal human eye in a solo field of view. However, the time base of the television raster allows for recording of any signal in the form of time-dependent analogue signals. In a sense, the signals can be either displayed or processed on a monitor with the similar time base. The signals can also be converted to computerized signals that can be either processed further or displayed.

The charge-coupled device as well as CMOS cameras have advantages over photographic recording. Only a few seconds are required by the charge-coupled device cameras to grasp a two- dimensional frame; it also has the potential ability to record a digitized image of the order of 10^7 or more pixel points. Furthermore, the response of charge-coupled cameras is linear for a choice of acquaintance dosage. Due to this, there is just a single relationship between the brightness that was observed and the signal's initial intensity coming from the object.

The produced computerized image can be processed more efficiently and easily by employing standard computer programs. The standard computer programs can aid in enhancing the image contrast by removing the background noise. The analysis can be done to extract the comparison information from the datasets in image.

Furthermore, the diameter of the charge-coupled cameras is 25 mm or fewer and therefore the image can be observed by the objective lenses solely. As such, requirement of any midway lenses is avoided. Major changes have taken place in the design of the optical microscope with manual control of the focus being replaced by computer control and photographic systems being replaced by charge-coupled device cameras.

1.4 Specimen preparation

Preparation of a good specimen is one of the major obstacles that are faced by many students to perform successful microscopy. The problems are idiosyncratic to the specimen. For instance, the sample response to grinding, polishing, and sectioning is determined by the elastic modulus and the rigidity of the material. On the other hand, the response towards chemical etching and electrolytic attack of the specimen is determined by its chemical activity.

The major problem associated with microscopes is that by focusing on the details, they tend to lose the larger picture. The relationship between a microscope image and the image obtained from an object can be lost track of easily.

Any engineering assembly is made up of complex constituents that have complex geometries. The constituents making up the engineering assemblies are often in different shapes and sizes. Various processing routes are employed to produce the different engineering constituents. As such they are unlikely to possess similar microstructure. The materials from which the constituents are made are nonhomogeneous and anisotropic. There is variation of chemical composition as well as the microstructural morphology across sections. The preferred orientation, for instance, aligned fibers, flattened grains, elongation, inclusion, etc., may be restricted to the microstructural morphology or the crystalline texture may be associated with it. In cylindrical texture there are certain directions that are aligned preferentially along certain directions in the constituent. Some of the examples of such cylindrical texture include the axis of the copper cable or the rolling path in the steel plate. Crystal-like microstructure might happen even in the absence of morphologic consistency, that is, the equiaxed appearance of grains is uniform though a common crystallographic axis is shared by all the grains.

Owing to the above discussion, it is clear that it is tough to choose how much the constituent needs to be sectioned to carry out microstructure examination. It is at all times convenient to describe the principal axes of the constituent and hence safeguard that the plane of any segment is allied with at least one of the defined principle axes. It is always necessary to inspect two segments even in the simplest of cases. The essential regularity axis of the element may be either parallel or perpendicular to the two sections. The sections for rolled sheets must be orthogonal to the three main directions – through-thickness direction, transverse direction, and rolling direction – in order to be desirable. In a big casting, the microstructure may change as a result of changes in cooling rate and the impact of segregation. The portions chosen for solidification from the initial portion of the casting and the segment obtained for solidification from the last portion of the casting differ significantly [2].

It is crucial to identify any other primary directions that are contained within the plane of section if the portion is orthogonal to the constituent's main direction. This might be the remnant of a free surface, the direction of rolling, or the direction of development. When mounting or preparing the slice, or due to the image inverting, either under the microscope or during processing, it is normal to become confused between structurally dissimilar orientations.

Many samples are required to be mounted for easy management during the process of surface readiness. The commonest of the sample holder may be either a polymeric resin or molding compound. This may be die-cast or hot-pressed without causing any distortion or damage to the sample. Sectioning of the sample may be required to fit it into the die hole. So, few numbers of samples can be supports in any positioning through the employability of coiled-spring or another instrument [2].

After the sample has been secured in its mounting position, the section of the surface can be ground flat and polished. Some care is required in the rough grinding process as it may result in excess removal of material, overheating of the sample, or introduction of damage (mechanical or thermal) at the specimen subsurface. The surface section must cut planar, before mounting the sample. There are different varieties of grinding media as such cemented, resin-bonded, or metal-bonded. Alumina, diamond, and silicon carbide are the commonest of the available media that are accessible in various grit sizes. The size of the sieve, which will gather the grit, defines the grit size and the quantity of apertures per inch determines the grit size. Grit size is a reverse function of the particle size. For instance, #320 grit has been composed by a sieve of #320 and has passed through the large, standard sieve size of #220. Sieving is no longer possible beyond #600 grit, owing to the aggregation of particles. For particle size corresponding to a few tenths of mm in diameter, a #80 grit could be used.

Grinding is a type of machining that uses the grinding medium's sharp edges to create cuts that are parallel to the sample's surface. The amount of material removed depends on the quantity of particle interactions and the shear velocity at the point where the grinding medium and workpiece meet. Heat produced at the contact and the debris left behind after the material removing process are the two main barriers to grinding. Heat created makes the workpiece more ductile, requiring more energy to eliminate the material. The resultant debris, on the other hand, jams the cutting edges. The heat production indicated above, along with the debris from the grinding zone can be reduced by flushing the surface being ground with coolant [2].

The blunting of grinding particles takes place throughout the grinding process. Wear of the surrounding matrix gives rise to new cutting surfaces and results in release of the blunted grit particles in the form of debris. This may occur naturally during the process of grinding workpiece. However, alumina dressing stone may be employed to dress the grinding tool. The choice of matrix is critical to accomplish the grinding process successfully. For sectioning of hard and brittle samples, metal-bonded grits can be employed. Resin-bonded discs are usually preferred in the grinding process because they are less likely to miss the cutting efficacy. Much coarser cut in sectioning operation is provided by the resin-bonded disc.

Direction of grinding is also important in sample preparation. For instance, care should be taken to not grind an area near a free surface that is orthogonal to the surface being ground. This is because the cutting particles may cause wide subsurface injury as these particles bite into the edge of the specimen being prepared. Damages will be much less by cutting in the opposite direction such that the grit elements exit from the free surface during grinding. The extent to which the subsurface of the specimen is being damaged depends to a large extent on hardness and elasticity of the material. It is difficult to grind soft materials. Brittle materials on the other hand are prone to subsurface cracking. Subsequent polishing ensures the minimization of the subsurface damage introduced as a result of grinding.

The primary objective of polish is to ready a surface that is plane as well as empty of topographical features. Each polished step is designed to eliminate the layer of dented subsurface layer form the previous stage of surface preparation. Electrochemical, chemical, and mechanical are the three means to polish a sample. Mechanical polishing has been proved to be the most important of all the processes. Finer and finer grit sizes are obtained through the mechanical grinding process that removes the mechanical damage taking place in the previous stage of specimen preparation. The number of steps that are required to reduce the topographical roughness to much lower than the wavelength of the light ranges from 3–10. A maximum of three steps of polishing is required in case of hard material whereas the number of steps can be a few tens for others. The carrier for polish grit may be paper-backing followed by a cloth polishing wheel.

On the other hand, in case of chemical polishing, a barrier of film is formed at the sample surface that inhibits the chemical outbreak. Thick layers of viscous and semi protective barrier films are formed on surfaces with negative curvature such as pits and grooves. On the other hand, the surfaces with positive curvature are more prone to chemical attack owing to the thin layer formed on the surface. Therefore, the surface is made topographically flat.

In the case of electrolytic polishing method, the surface is required to be an electrical conductor and therefore it is only preferred for metallic surfaces. Positively charged cations are dissolved in the electrolyte, the reaction taking place at the surface of the specimen. As a result of the reaction, a viscid anodic-film of large resistance is formed at the surface with higher electrical resistance. To achieve effective electrolytic polishing, the majority of the voltage drop should occur across the anodic film. This is because, as in the case of chemical polishing, the film thickness controls the rate of chemical attack. Better control for electrolytic polishing in comparison to the chemical polishing can be accomplished by making external adjustments to the voltage across the electropolishing cells. Moreover, soft materials that are difficult to be polished using mechanical grinding can be ground using electrolytic polishing. Recent advancements have resulted in the growth of highly cultured mechanical polish approaches that have potentially replaced the chemical and electrochemical means of carrying out surface readiness.

Engraving removes material from the surface in a choosy manner and develops features on the surface that are associated with the microstructure of bulk material. Etching, however, may be unnecessary in case the different phases reflect and absorb the light in a differential manner. In case of most nonmetals, etching may be not required owing to the visibility of the nonmetallic inclusions in the resulting engineering alloys. This is because most of the light is reflected by the metallic matrix while the absorption of light takes place at the inclusions. Optically, anisotropic samples also reveal contrast without etching and are associated with the differences in the structure of the crystals.

Surface topography may also be developed with the help of the etching process. Grooving the grain boundaries or the differences in height of grains present in the neighboring surfaces are few examples of development of surface topography. Thin surface film development is another feature that can be developed through the etching process. The thickness of the surface films reflects the grain structure and the underlying phase. Presence of such thin films on the surface may either absorb the light incident onto the surface or may give rise to interference effect. This depends on the thickness of the film.

Chemical attack is normally involved in most of the etching methods. This is more pronounced in the surface regions that have higher energy, as for instance the grain boundaries. Thermal etching is, however, an exception. Short-range diffusion of the surface takes place in case of thermal etching that involves the heating of sample. The diffusion process reduces the energy of the polished surface situated near the boundaries. As such, the local topography is affected to a large extent. Thermal grooves are formed at the grain boundaries. In many cases, the surface energy associated with many materials is highly anisotropic. As a result, the thermal energy of the surface is reduced owing to the formation of surface facets in case of some of the aptly oriented grains.

Chemically active solutions are usually employed to accomplish etching in most cases. This is the most common method of etching to reveal and develop the surface topology under the microscope. The solvents employed are usually alcohols; however, in certain cases such as that of reactive samples, molten salt baths may be used. Particularly for ceramic specimens, molten salt bath may be employed. In most of the cases the samples may be dissolved in either of the aforementioned solutions for a given time and at controlled temperatures. These are then rinsed thoroughly using alcoholic solutions. In some cases, such as stainless steel, electrolytic etching may be employed to aid in promotion of localized attacks.

Surface films may be sometimes required to be formed on the surface and this may be accomplished through chemical staining. However, the surface characteristics of the microscope have an impact on the film's thickness. Such surface coatings will form on an oxidized steel sample in the presence of air. A spectrum of interference colors will be present among the various grains. The temperature and oxidation time regulate this. The coherent oxide coating thickness that develops is determined by this on each grain.

The most common of structural materials are the engineering alloys that are ready for microscopic examination by the process of mechanical polish. There are certain problems associated with the mechanical polishing process that are yet to be addressed. Surface relief is the most common of the surface defects arising due to polishing. Plastic deformation, scratches, and rounding of edges are some of the other defects associated with mechanical polishing. If it is assumed that the polish media permits dirt, then the defects such as scratches and plastic distortion may be minimized through proper selection of the compliant support. The presence of compliant support reduces the force applied to the distinct elements and also enhances the number of particles that make unit area contact with the surface. A less compliant support on the other hand aids in inhibition of edge rounding and surface relief.

It is hard to polish soft materials through the means of mechanical polishing. The wear debris will get surrounded by the sample surface in case extreme strength is applied to the polish media. The embedded wear debris will then be dragged in the direction of shear. Even in cases where the damage signs are invisible, subsurface traces associated with plastic deformation are revealed through the subsequent etching process. Polishing at low shear velocity results in successful preparation of such materials. This process prevents the debris from adhering to the sample being prepared.

It is easier to polish brittle materials by mechanical processes in comparison to soft materials. In particular, the issues related with the debris stick to the surface of the specimen are less likely. However, there are other imminent problems as such microcracking and grain pull out that may arise during mechanical polishing of such materials. Such defects are mainly associated with the adhesive failure at boundaries and interfaces.

Therefore, selection of a suitable polishing medium is essential. The hardness of the polishing medium must exceed that of the sample under preparation. Neither silicon carbide nor alumina can be polished successfully using SiC grit and in such cases diamond is the preferred polishing medium. The same is true for the silicon nitride samples. There are some advantages associated with cubic boron nitride as polishing medium. The hardness of cubic boron nitride is more than SiC and the oxidation resistance is superior to diamond.

The fact that several of the most brittle materials may bend flexibly in the high-pressure region is an important issue worth mentioning, especially the high-pressure area underneath the spot where the grit particle comes into contact. Internal stresses will be generated in the region of plastic flow and this will result in crack and chipping about the unique contact once these stresses are removed. Once more, preventing the abovementioned issue requires limiting the pressure that is applied. Another way to limit the abovementioned problem is to use a grinding media carrier with a higher compliance.

Engineering composites are the most problematic materials to formulate for investigation through optical microscope. In case of composite materials, a soft compliant matrix is strengthened with a brittle fiber. Although, composite materials got equipped for optical microscopy, few studies have been published depicting dispersal of the strengthening parallel to the fiber. Preparation of such samples has resulted in loss of fiber adhesion, support to the matrix by fragments of hard reinforcing agents, fracture of loose fibers, etc.

With an understanding of the problems and techniques involved in sample preparation, it is also important to identify samples that may be just inappropriate for carrying out optical microscopy.

1.5 Image contrast

There are several ways to develop image contrast in the optical microscope. Most of the methods for image contrast require careful sample preparation. Quantitatively image contrast can be defined as the intensity alteration among neighboring resolve features, $C = ln\ (I_1/I_2)$. However, in case of small difference in intensity, the image contrast is defined as $\Delta C = \Delta I/I$. It is suggested that ΔC is at the smallest, 0.14, if features detached by a span equivalent to the resolve are observable.

At large partings, very small contrast dissimilarities can be distinguished; there are several image analysis set-ups that function with ΔC equivalent to 0.004, as well. Objective lenses with higher values of numerical apertures have the potential capability to accept scattered light at varied angles in comparison to the objectives with lower values of numerical aperture. Therefore at higher values of numerical aperture, the contrast of the image from any specified features is lowered. Also, at higher magnifications, images tend to depict lower contrast.

Electromagnetic radiations incident on to the surface of a polished surface might be absorbed, conveyed, or reflected back. Specular surfaces reflect the electromagnetic radiations to a great extent and are significant for the existence of free conducting electrons in the specimen under observation. This is reflective of metal material, though many metals trap most of incidence radiations. For instance, metals as such gold and copper absorb the blue spectrum of light and as such the reflected light appears yellow or reddish. On the contrary, metals such as aluminum and silver reflect 90% of the incident radiations and as such are employed for mirror surfaces.

The high reflectivity possessed by polish aluminum has no effect on the inclusion of amorphous oxide protected layer because the thickness of the deposited oxide film onto the aluminum surface is much lower in comparison to the wavelength of the visible light. Indeed it is worth noting that the metallic surfaces are covered by a thin layer either as a consequence of sample creation or as a result of atmospheric reaction. The deposited film does not interfere as long as the film is uniform and coherent, with a thickness lower than that of the wavelength of the visible light.

The angle of the incident light determines the relationship between the fraction of the light incident and that absorbed, transmitted, or reflected. The critical angle of light of incidence beyond which light cannot be transmitted is dependent on the refractive index of the solid. The critical angle can be exceeded in dark-field illumination and this may result in light indicator scattering into the objective lens. The thickness of the sample does not significantly affect the portion of the light incidence reflected from the surface given the condition that the thickness exceeds the wavelength of the incidence light. However, the fraction of light reflected from the surface is dependent on the material of the specimen as well as the angle of incidence. It is a fraction of light transmitted from the surface, which depends on the thickness of the specimen. This decreases exponentially with the thickness of the sample [2].

Mineralogical samples are ready as thin segments and are inspected in the presence of polarized light. It is necessary to ensure that the samples are thin so that adequate transmission of the incident light may be allowed. Adequate thinness of the sample also ensures that the resolved features of the specimen are not overlapping with the projected image of the slice. Typically, a sample thickness of 50 μm or even less is suitable.

In the case of ceramics and polymer samples, poor contrast of images is obtained as such specimens convey or trap a considerable portion of the incidence light. This may be attributed to the weak reflection of the signal and also because of the scattering of the light that takes into the objectives from the subsurface feature of specimen that are beneath the focal plane in the image. An evaporated or sputtered coating of aluminum may aid in highlighting the surface features, but may result in loss in information that is associated with the variation in absorption and reflectivity. Some of these specimens may be comfortably studied when conveyed through the polarized light.

In case of normal, bright field illumine, only a portion of the light that is incident on to the specimen is scattered back to the objective lens. The brightfield illumination comes with two limiting contrast options. The first option is the one that has been debated earlier and is referred to as Kohler illumination. The incident light is focused at the back focal plane of the objectives. As such the light is incident normally onto the sample from a point source and the light from a stretched surface is incident over a scope of angles that is determined by the dimensions associated with the source image in the back focal plane of the objective lens. If the source of the light is imaged onto the sample plane, then the dispersal of the light concentration over the surface of the specimen will replicate that in the source of the light. The image of the source of light will be visible into the microscope, if the specimen is a specular reflector. The top excellent light sources produce the light evenly and due to advantages in enhancing the concentration of the incident radiation by concentrating the condenser arrangement so that there is coincidence between the planes of source and the plane of the samples. However, in most of the cases, especially at low magnification, focusing of the light at the back of focal plane will aid in uniform illumination of the sample.

The image of the topographical features that lies outside the realm of the incident light appears to be dusky in the image. This is particularly right in cases of steps and grooves present in the grain boundaries. The dimensions of such features may even be fewer than the limiting resolve of the objective. The two characteristics will only be visible as image if the distance between the two features is greater than that of the Rayleigh resolution. In such cases, sufficient contrast in the image may be produced. In most of the cases, the size associated with the topologic characteristics at the surface mirrors the surface readiness in the same proportion as that of the bulk microstructure. Grooves at the grain boundaries as well as the phase boundaries are the prominent examples of such a case and are associated with the thermal and chemical itching. It is always suggested that considerable care needs to be taken during the measurement of porosity, the particle size, or any other microstructural parameter.

The contrast is chiefly resolved by making the comparison between the concentration of light from the source and the background. As such, darkfield objective can be employed to enhance the contrast of the features that appear to be dark in normal, brightfield image. Deflection of condenser systems may be required in certain cases to enhance the topographical information. In such cases the sample is illumined only from a single side. Such slanted brightness is referred to as being available as standard attachment. The three-dimensional features are revealed by seeming shadowing of the characteristics by using slanted brightness. However, this process may result in the loss of information. There are certain cases, as mentioned previously, wherein the images produced may be somewhat misleading. This is because the seeming recognition of the crest and trough on to surface depends on direction of the incident light. If the sample is illuminated from the top and the microscope is arrange to illuminate the sample from below, then the valleys will appear to be hills.

The study of transparent polymers and glasses may be conducted through confocal microscopy that has been established chiefly for biotic and healthiness. A parallel ray of light is taken down to a sharp focus in confocal microscopy. Laser is the main light source for confocal microscopy; however, in some cases, monochromatic green light from mercury and white light from xenon arc have been employed fruitfully. The point probe is scanned transversely; the sample and signals are generated from the thin slice of specimen that is situated at a definite deepness below the surface. The resultant image is referred to as optical section. A sequence of optical sections that are taken at changed zeniths below the surface is revealed to be allowed from background noise. A very high-resolution image in three-dimensional form of some definite locations in the biotic tissue sample have been produced by employing a focused cone of light and through suitable selection of the fluorescent signal [2].

In the case of interference microscopy, there is interference of the light reflected from the sample with that of the light from an optically smooth typical reference surface. The two rays must be ordered, that is, they should be in immovable relation and this can be ensured if both the beams are made to originate from the same source. This is done via a beam splitter that primarily separates and then recombines the two signals. One of the simplest ways of achieving interference contrast is to use a layer of aluminum or silver coating to cover an optical glass coverslip. This arrangement ensures that more than half of the unicolor incidence light beam is communicated by the cover-slip. The light that is reflected from the thin metallic layer is known as reference ray, whereas that conveyed from the reference covering and then from the sample surface is known as the interfering beam. It is also expected that there is no assimilation of the light and therefore the incidence light is either transmitted or reflected and no absorption of the incident light takes place [2].

Say, R is the reflection constant of metallic layer, then the transmission constant will be $(1-R)$ in case there is no absorption of the incident light. If 1 is the reflection constant of the sample specimen, then no light is captivated and the intensity of the light reflected back from the sample will be $(1-R)$. The metal sheet will then send the sample surface's reflected beam back, with a transmission coefficient of $(1-R)$. The light's residual intensity will be reflected back. As a result, $(1-R^2)$ will be the coefficient for the intensity of the second light source from the sample, which is transmitted backward over the metal coat. Since the two rays must be of identical intensity in order to interact severely, R is equal to 0.38 [2].

The two beams must have a path difference of at least $(2n + 1)/2$ for harmful interference to occur. As a result, there will be a phase difference of between the two beams. On the other hand, the path divergence is just twice as far apart as the reference surface that is partially reflecting and therefore the following equation holds true $2h = (2n + 1) \lambda/2$. When the two surfaces are separated by a distance equivalent to $\lambda/4$ then destructive interference occurs. Successive interference fringes are produced that are separated by height $\Delta h = \lambda/2$. An easy two-beam interferometer will be skilled of topologic height changes to the precision of ±20 nm given that the moves in the interference fringe are measurable to the command of 10% of fringe parting and also the wavelength of the incidence light is in green.

Placing a droplet of immersion oil between the sample and the coverslip results in improved sensitivity and contrast. The effective wavelength is reduced by a factor μ, that is, the same as the refractive index of the immersion oil. The numerical aperture of the objective lens is limited to cause a good interference between the beams. This is because if the numerical aperture is kept greater than 0.3, the change in the path length of the light transitory over sideline of the lens can result in destruction of the coherency of the beam. To safeguard that the incident light is normal to the sample surface, the condenser setup could be focused on the focal plane at the back of the objective lens. In certain cases, it may be of great information to view the sample in white light and this is valid when the Newton's colors are exhibited by the orders of the interference fringes. Complementary colors result owing to the deduction of single wavelength from the white spectrum. In the case of very small separations, the interference of the shorter wavelength incident light takes place in blue and the corresponding Newton's color is yellow. As the parting surges, the wavelengths interfering move to the green and the resulting corresponding color is magenta. As these progress, the interference changes into red and the resulting corresponding color is cyan. For still higher orders of partings, the interference condition moves to still higher order and the sequence of colors obtained is repeated. However, the contrast produced is dull because lights with dissimilar wavelengths begin to accord towards interference [2].

A white appearance in the region of the sample that is in interaction with the reference plate is observed. It can be observed that there are a few points of contact amongst the reference plate and the plate even when the specimen is flat and sample is polished properly.

In the systems for interference microscopy, the path differences that might be introduced by the optical system may be avoided by the employability of two objective lenses. The reference surface and its position must be maintained lengthways; the optic axis and reference surface can be arranged to be slanted about two perpendicular axes. The reference surface can be tilted in the plane of the surface and this and the aforementioned arrangements can ensure adjustment of the orientation and the spatial separation of the interfering fringes.

In order to ensure that Newton's interference colors are observed, the location of the reference outward is attuned at a small angle tilt in the white light. At this position, the white light marking the line that is coincident with the reference and the specimen image appears to be observed in Newton's interference. Simple systems for interfering microscopy are accessible as extras for interference microscopy, but most of the available attachments have limited life. This is because most of the systems rely on half-silvered reflecting surface and as such the surfaces are damaged easily once they are brought in contact with the specimen. The incapacity to regulate the parting between the sample and the reference surface or the distance between the fringes is one of the disadvantages associated with a metal-coated coverslip. However, this is not critical for many applications.

In the case of a simple two-beam interference, a part of the light is reflected simply. There are intensity losses to the tune of 24% for the aforementioned case wherein the assumptions of no loss associated with the absorption and reflection constant of 0.38 for the half-silvered coverslip have been considered. With the losses associated with the intensity of light amounting to 24%, unwanted background is reflected in the images. The proportion of the light being reflected multiple times can be increased with the increasing reflection constant of the reference surface. The value of reflection coefficient can be increased up to 1. The intensity of light reflected multiple times can be determined using eq. (1.7):

$$I = \left[\frac{T^2}{(1-R^2)} \right] \left[\frac{1}{1 + \frac{4R}{(1-R^2)\sin^2\left(\frac{\delta}{2}\right)}} \right] \tag{1.7}$$

where T and R represent, respectively, the transmission and the reflection constants. The reflection constant of the reflection surface and that of the sample under consideration can be assumed to be equal to 1. Parameter δ, depicted in eq. (1.7) can be determined using eq. (1.8):

$$\frac{\delta}{2} = \frac{(2\pi h \cos \theta)}{\lambda} \tag{1.8}$$

where the parting of the reference surfaces and the sample surface is represented by h. the collected intensity falls to zero when $2h \cos\theta = n\lambda$ and it is assumed that there is no assimilation in the reference film, that is, $T + R = 1$. The interference fringes formed are restricted to the reference surface and it is the angular slant of the reference surface with respect to the optic axis and the distance among the reference surface and the sample that determines the number of beams contributing. The number of beams taking part in the interference determines the width of the black interfering fringes [2].

The incident beam is required to be parallel and the surfaces are required to be separated by not more than a few wavelengths for effective multiple beam interferometry. Spin-coating of the samples with the aid of slim plastic film and evaporating silver onto the smooth surface are some means that can ensure best patterns. Topological images at nanometer levels can be revealed by very sharp interference fringes under such conditions. Moreover, it is probable to image the development stages associated with crystals that are characteristic of only limited atoms in height.

1.6 Working with computerized images

The terminologies associated with the computerized imaging are discussed in this section. The discussions made are quite general and apply equally to the images obtained from optical or electron microscope and also that derived using raster scans.

1.6.1 Data collection

The intensity of the image obtained from a simple optical microscope is a continuous function of the coordinates in the image plane, that is, if the image plane is x-y plane then $I = f(x, y)$. The number of electrons that contribute to image formation is large and as such ignores the statistic associated with the formation of the image. However, this is not the same with the damage-sensitive materials. On the other hand, the signal-to-noise ratio may be an issue as in the case of weak signals. The signal is reliant on the probe energy, that is, the wavelength of the light employed and therefore one can use the following relation: $I_\lambda = f_\lambda(x, y)$. The aforementioned equation is valid for equivalent optical images gained in transmission electron microscope and monochromatic light. In case of the color image recording optical microscope depends on the method of sieving the intensity over the color filter. The colored filter may be red, green, or blue (RBG) for positive images and for negative images this may be magenta and cyan. The equivalent intensity, therefore, is reliant on three color parameters and also on the two spatial constraints. Only one wavelength is involved in case of images noted in unicolor light. However, the wavelength still needs to be specified [2].

1.6.2 Data processing and its analysis

The variation of intensity with the position, that is, $I = f(x, y)$ signifies the info of the item transmitted by the radiations passing over the optical system. The intensity of the image is a continuous function in x and y. The intensity function is, however, modified owing to the response of the recording medium in case the recording has been accomplished in analogue data format. However, in this case too, the intensity function is a continuous one [2].

In the case of computerized data format, the intensity function is not continuous function on the x-y plane. The intensity function is a discrete function. The image is stored in the form of a digitized setup as distinct image elements. These are also known as pixels and every pixel provisions the info on intensity associated with the image at every x and y location.

The dimension of a pixel is identified by the entire zone of the image noted and the whole number of pixels. The area tested by a definite pixel and the location on the sample under consideration are associated equivalently with one another. The two important factors that are required to be considered are: resolution of the microscope and efficiency of signal collection. The pixel dimensions are to be kept lesser than the resolution projected on the image plane so that full resolution of the microscope can be assured. In usual practice the dimension of a pixel is kept at 3 × 3, so that the features represented are resolvable. This dimension of pixels also confirms that no resolution is misplaced in the computerized image.

Recording of the image always takes place in a rectangular format. This assures that the number of pixels are not same in x and y axes. A ratio of 4:3 is the recommended aspect ratio for television and video equipment. On the other hand for CCTV cameras it is 1:1 and HDTV has an aspect ratio of 16:9. It is often observed that the image recorded in one format appears to be distorted when taken to other format.

The x and y axes magnification rarely match for the SEM. There may also be variation of magnification from the center to the periphery of the view. Depending on the size of magnification, a square grid appears either to be pin-cushion or barreled. The size of magnification near the center of the scanned area is an important consideration. Foreshortening perpendicular to the tilt axis occurs in case the sample is tilted without loss in focus in the SEM. Different magnifications in x and y directions can be obtained if calibration work has not been carried out. Therefore, calibration or correction of the work being carried out is one of the important considerations to ensure uniform magnification in x and y axes.

A person's eye only uses all of its resolution on a small portion of the retina. The eye is capable of scanning over a distance of 20 cm at the close point. The least number of pixels to be noted for a computerized image is of the order of 10^6. In case of dynamic imaging, the human eye can comfortably keep track of the deviations in the image and somewhat lesser pixels are required in each frame. While in case of the dynamic color images, the computerized image depends on the dyes to choose the RGB constituents in neighboring pixels. Characteristically this is accomplished in a 2 × 2 collection of four pixels. In such an arrangement, the two computerized constituents are red and blue while the other constituents are both green [2].

1.6.3 Storage of data and its presentation

So far the discussion has been regarding signal digitization and no discussion has been made on digitization of the recorded intensity from a given location or from a pixel. One needs to be familiar with the terminologies being used. The indication developed from a pixel at a given location characterizes the intensity. Intensity per unit area is known as the brightness. As such brightness can be defined as the summation of the intensity over the area or the pixels defining the area and the summation divided by the area covered by the pixels. Sampling of the pixels at the spatial intervals aids in identification of the differences in contrast. These are smaller in comparison to the distances separating the variations in intensity. As such the high density of pixels will aid in preservation of the spatial frequencies in the unique image. The original image contains the associated information. As per the Nyquist criterion, the interval of pixel sampling should be double the peak spatial frequency to be recorded. On the contrary, as per the Shannon's sampling theory, the sample recess should not be more than one-half of the resolution limit of the optical microscope setup. For all practical aspects, the aforementioned standards are alike. However, it is worth noting that there will be a noise boundary that limits the consequence of the alterations in the contrast.

There are two separate factors that can bind the valuable scope of concentration that can be noticed by the in-place recording system. The primary limit is the minor limit that is due to the background noise or by the letdown of the arrangement to hold indicator with low intensity. The secondary limit is the greater limit that is set by the fullness of the recording system or the detection setup.

In the case of photography, the following are the two limiting conditions: reciprocity failure at low doses and the overlap associated with the precipitation corresponding to high dose. Moreover, the response curve associated with a blend is best conspired on a curvature of density and the log of dose. The linear portion of the curve plotted is the useful range of curve. The contrast response is determined by the slope of the curve.

The analogous curves obtained in the computerized records are alike, but in such instance the signal noise determines the low limit while on the other hand the saturation and the blooming of detector reflect the high limit. There are two systems that are common in usage. The electric charge excited by the image signal is collected by the CCD detector. The collected charge is read off from every pixel when the exposure time gets completed. Pixel sizes may range from some micrometers to tens of micrometers. Moreover, the size of the frame, that is the whole quantity of pixels in the array, is also flexible. A distinctive dimension of the frame may be 20 mm transversely and may contain 10^7 pixels or more.

Complementary Metal-Oxide-Semiconductor (CMOS) detector is the second computerized image set-up. Instead of reading the pixel signals as an entire data frame, it is read line by line. The read-out circuitry is located around the rectangular array of the CMOS system for computerized recording, which is essentially a camera on a single chip. Both of the aforesaid detectors are used for electrical data collecting and optical picture recording. However, CMOS technology is a far better choice than Charge-Coupled Device (CCD) technology in situations where dynamic real-time recording is required. The CCD technology is employed in the case of electron microscopy. Over the detectors, dye filter arrays are used to provide colored photos.

1.6.4 Computerized storage

The human eye is capable of distinguishing 20 intensity stages in a grayscale image. Also, a wide scope of color shades and tints are distinguishable by the human eye.

However, the color response is totally different for two different individuals. Also it is impossible to record the image as perceived in nature. In case of recording a computerized colored image, the task seems to be almost impossible. What a human eye sees over the eyepiece of an optical microscope appears to be more vivacious in comparison to the image that can be recorded or that can be recorded on a colored computerized display. The positive color print is an even poorly recorded version of what is perceived through the eyepiece of microscope. Also, in each case, the tonal values will also be dissimilar and will differ from one recording setup to another. Also, there is variation obtained with the settings by the operator.

The range of tints and shades, in case of both the analogue and computerized image color observation, is limited by the presentation of the pigments, dyes, as well as phosphors. For a clear description, a grayscale image is considered. A black as well as white image is stated as a 1-bit binary image. Every pixel in the image is either recorded as either black or white. It would be 2-bit image in case four levels of intensity are noted per pixel and 2^n gray-levels per pixel can be kept in n bits of image data. Most of the setups can store more than 4 or 5 bits that are required to keep all the gray levels that have been detected by the human eye [2].

A great deal of information can be added through color to the image being recorded. In such cases 8-bit data is stored that corresponds to 256 intensity levels in each of the colors in blue, green, and red. It is, however, needed to work the levels of intensity when processing the image. This ensures full use of the scope of intensity level that must be noticed by the eyes. This is, however, true for both the cases of grayscale and color computerized images. There are numerous formats in which the computerized data can be stored and made available. A few instances include JPEG ("Joint Photographic Experts Group"), BMP ("Bitmap"), TIFF ("tagged image file format"), etc. All the formats differ from one another in the algorithms used by the formats to store the computerized data. Microscopists employ the format as per their convenience. All the available data formats will compress the data unless the compression option is disabled throughout the file storage. The compression of large files is convenient only while transferring or storing files. However, during the compression loss in data always takes place.

A brief discussion is now made on printing of computerized images. Most of the printers that are available print a dot-matrix array and are halftone. The tones, however, depend on the density of dots per unit area of the page being printed. There are dual types of halftone printer that are available for general use. Inkjet printers are those in which a piezoelectric sensor ejects a droplet stream from a reservoir containing black or colored ink. The colored dyes may be red, blue, green, cyan, magenta, or yellow. The droplet of ink is captured and dried on the paper with insignificant dispersal. On the other hand, laser printers adopt a totally different technology and create a charge design on a photoelectric surface. Laser printers are fast in comparison to the inkjet printers, but are expensive. In both the cases of printers, it is the cost of dyes and toners that determines the economics associated with the procedure.

1.7 Image interpretation

Some of the issues related with the understanding of optical images obtained from microscopy have been discussed now. The noted image of a sample will contain information associated with the microstructure of the material if the image is processed through the following sequence of processes: (1) sample preparation, (2) imaging in microscope, and (3) observing the recorded image. The contribution from the above phases will determine the final resolution of the characteristics observed in the noted image. The resolution of the fine-grained microstructure may be limited by the etching process. Etching affects the resolution far more than the numerical aperture of the objective lens. However, this may also be affected by the grain size of the used emulsion for recoding of the images. The photographic process adopted may also influence the resolution. In the case of computerized imaging, the restrictions on the data processed and resolution are determined by the size of the pixel, the number of pixels accessible, and the number of gray levels employed to record the brightness of the image.

Surface topology, the existence of surface film, and the optical properties associated with the sample are the influential factors for the contrast in the optical microscope. The observed contrast is reliant on the wavelength of the light and might be related with the amplitude differences in the reflected light. This may also result from phase shift as in the color observed with a sensitive tint plate. Resolution is itself of no use in images without contrast. On the other hand, poor resolution just for obtaining a decent contrast is correspondingly detrimental. The eye is the best judge of what establishes a good negotiation. CAPP, that is, "computer-assisted data processing" can, however, offer better, unfailing, and objective support [2].

1.8 Illustrative example on characterization of nanocomposites through optical microscopy

Polarizing microscopy is a valuable technique for the examination of food products that display optical anisotropy or birefringence. Numerous components in food, such as starch granules, fat crystals, and protein fibers, exhibit birefringence. When it comes to starch birefringence, polarized microscopy is particularly effective. This is evident from the distinct and high-contrast 'Maltese cross' patterns observed in unheated specimens (refer to Figure. 1.1). Determining the gelatinization temperature of both native and modified starches involves monitoring the progressive loss of birefringence as the sample is heated and gelatinization occurs. Moreover, polarized light microscopy is also a valuable tool for characterizing the morphology and size distribution of fat crystals. These characteristics are closely linked to the stability and quality attributes of lipid-based products (as illustrated in Figure. 1.1), owing to the birefringence exhibited by crystallized fats.

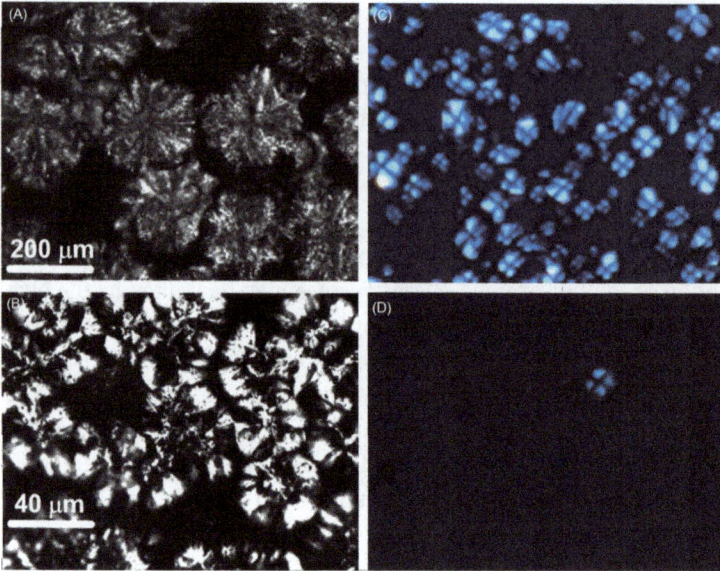

Fig. 1.1: Polarized light micrographs of cocoa butter with two solid fat contents: (A) 20% and (B) 70% and of starch granules (C) prior to heat treatment (D) after heat treatment (right) (reproduced with permission from [3]).

Polarized OM was employed to study the cocoa butter with two solid fat contents by Maria G Corradini and D Julian McClements [2]. Certain food constituents exhibit birefringence, with notable examples being starch granules, fat crystals, and protein fibers. When employing polarized microscopy, starch birefringence can be easily distinguished by observing distinct and high-contrast "Maltese cross" patterns in specimens that have not undergone heat treatment. This has been depicted in the OM images in Fig. 1.1.

1.9 Conclusion

Optical microscope is the preferred tool for carrying out microstructural characterization of the engineering materials. This may be attributed to the wealth of information that is provided by the amplified image and also the ready accessibility of such instruments. Moreover, the specimen preparation facilities are also available readily and the techniques associated with the processing and recording of data are widely available. The visual impact produced by the image is instantaneous and its understanding can be made in terms of spatial relations and hence are comprehendible to the observer of the macroscopic world.

The relation between the object located in the microscopic step and the image produced can be determined with the conceptual framework provided by geometric optics. The wavelength of the light that images the object limits the resolution detail that can be provided by the optical microscope. Also, the limitation on the resolution can be placed by the angle subtended by a point in the object plane at the objective aperture. The range of wavelengths associated with the light and the aperture of the eye can limit the resolution of the eye to nearly 0.2 mm. This is significant for the fact that the two features that are distanced 30 cm can be distinguished by the eye if the features are separated by a distance of 0.2 mm.

The numerical aperture of the objective lens is given by $\mu sin\alpha$, where the refractive index of the medium is given by μ and α is the angle subtended by the object being imaged at the objective lens. The values of numerical aperture vary from 0.15 for objective lens with low magnification power to 1.3 for an objective lens with high power, oil immersion objective lens.

Chapter 2
Transmission electron microscopy

2.1 Introduction

There are certain similarities between a transmission electron microscope and an optical microscope. However, the transmission microscope is inverted in logic in that the electron origin is situated atop the microscope and the recording medium is at the bottom. The optical source of light is replaced with the electron gun, and the electrical energy is typically preserved higher in comparison to the ground. A number of different sources of electron rays have been developed. However, the construction principle in these microscopes with varying origins of electron ray is similar. The electron is extracted from a heat filament at little prejudice to the electrical energy from a thermionic source; a low voltage being applied between the cylindrical polished cap and the source. An electrostatic field aids in focusing the ray of thermionic electrons. These are then accelerated by the anode that is seized at ground potential below the Wehnelt cylinder. The parameters characterizing the electronic ray entering the microscopic column are the operative source dimensions d, the deviation angle of the ray α_0, energy spread of the electron ray ΔE, and the energy of the electrons E_0. The ray coherence is improved by employing a small source size d and the contrasts are obtained through the phase shifts, owing to the exchanges of the ray when it passes through the sample.

The three common sources of an electronic ray are: heated tungsten filament, lanthanum hexaboride crystal, and Cerium hexaboride. The electron ray densities from the tungsten filament are of the directive of 10^6 A m^{-2}, from an operative source size which is determined by the first crossing of the electron ray, and is typically 50 μm. The energy extent is 3 eV, owing to the high thermionic emission temperatures. Furthermore, the coherence nature of the ray is also inadequate. A higher ray current, of the directive of 10^6 A m^{-2}, is generated in case of lanthanum hexaboride crystal, but at relatively lower temperatures. In the case of cerium hexaboride, the energy spread is reduced to about 1.5 eV. However, in this case, the requirements on vacuum are stringent in comparison to the tungsten filament electron source. Electron bulks, of the directive of 10^{10} A m^{-2}, have been achieved through the incorporation of a cold-filed emission gun, wherein the electrons channel out from a shrill point below the effect of an electric field of great intensity. The diameter of the sharp tip is no more than 1 μm. As such, the effective size of the source is no more than 0.01 μm, and hence highly coherent. The temperatures generated are low and the total energy distribution is 0.3 eV. The cold source r is replaced by the hot field emission basis. Hence, the tungsten needle-shaped filament gets heat to improve the process of electron tunneling, and the procedure is termed as Schottky release. The work function is reduced with zirconium in Hot Schottky sources. However, they possess better energy feast

https://doi.org/10.1515/9783110997590-002

and a greater operative basis dimension in comparison to the cold-filed emission sources. Such sources are additionally reliable as well as stable and have a longer life as well as smaller requirement of vacuum. The energy spread from the electronic source is reduced to less than 0.15 eV through the overview of electron ray monochromators. However, the reduction in energy feast occurs at the expense of attainable electron current mass. The decrease in energy feast becomes important for analytical analysis, and also aids in improving the info boundary in the transmission electron microscope.

The high-energy electrons from the cannon are concentrated by the employment of an electromagnetic condenser lens system. The control in the lens current helps in adjusting the focus. The specimen stage has mechanical complexity. In addition to the control of the specimen in the x-y plane, the specimen can be slanted about 2 axes perpendicular to the plane of sample. Therefore, adjustment is also required with respect to the tilt axes. Some adjustment for the z direction is also important. The specimen can also be rotated about the optical axis of the transmission electron microscope. The sample diameter in the case of transmission electron microscope is only 3 mm. Moreover, samples of size less than 0.1 μm allow most of the electrons to pass through them without any loss in energy. While in the case of SEM, a contrast is gathered from the sample surface and there is loss in the energy of the of electron ray as the electrons penetrate the surface of the sample. The concentrating of the image in the case of transmission electron microscope is not accomplished by positioning the sample lengthways along the z-axis but rather by adjusting the lens current that controls the focal span of the electrically magnetic lens. The lens current is adjusted so as to first produce an image from the elastically dispersed electrons that have been conveyed over the thin film sample. Electromagnetic lenses are also employed in the final imaging process wherein the ending image is detected on the fluorescent display, which alters the image produced by the high electron energy to the image that can be perceived easily by the eye. The characteristic electron current bulks at the display are in the range of 10^{-10}–10^{-11} A/m^{-2}. However, this might be lesser once studies are carried out for damaged sensitive resources or while taking images at higher amplifications. Photographic emulsions have been developed to obtain ending images; however, charge-coupled devices have been developed that combine with computerized image processing.

There is a limited path available in the air for the high-energy electron ray to travel and therefore the microscope pillar has to be necessarily held beneath the void. A growth of carbonaceous deposit over the sample surface is known as specimen contamination and is one of the serious problems in the case of electron microscopy. This might limit the watching period for any particular zone of the specimen and firmness may be caused. As such, vacuum of more than 10^{-6} Torr is needed and maximum firmness is achieved for a void of 10^{-7} Torr. The source of adulteration may be the pumping system, constituents of microscopy, or the sample, and these can be avoided by cryogenic chilling of the sample and its environments.

In the case of scanning electron microscope, the major components are the source of high-intensity electron ray and the condensing arrangement. It is presently used like a probe lens that focusses the electron ray into a well probe and then penetrates the sample. The function of the objective lens in the transmission electron microscopy is served by the electromagnetic lens system in the scanning electron microscope. The probe lens is, however, positioned over the sample, which does not have any role in gathering the gesture from the sample. In SEM, no morphological-associated information is provided by the elastically scattered electrons. Rather, the microstructural information is provided by the scattering of inelastic electrons that happens when the electron probe relates the specimen. The energy of the electron ray, employed in the case of SEM, is lower in comparison to that used in transmission electron microscopy. The energy of the electron ray usually ranges from 2–30 keV, and even much lower energies i.e., 100 eV may be useful.

Focusing the electron probe across the specimen, results in the formation of the image in SEM. The image signals are then collected, which go through suitable magnification and processing. These are then displayed on a screen with the period base similar as that employed to crosswise scan the specimen by the probe. At first, line scans are made of the sample along x-axis and then at the termination of every stroke scan, the ray is swapped back to nil of the x, and then the y coordinate is increase by Δy. The collected sign is a function of period t i.e., $I(t)$ wherein every value of t is associated with that of the x, y coordinates in the 2-D of the sample. After the signals from the line scans have been collected, the ray is swapped back to nil in the x-y plane. Once swapped back to the nil coordinate, the probe is prepared to gather information from the other image frame. A similar principle is used in the CRT i.e., "cathode ray tube", where the image is formed by scanning, employing a controlled electron ray. In this case, the increment is done simultaneously in x and y and the total time taken to scan the entire image frame is given by the whole numeral of points in the times of multiple frames with the residence period at every location of the pixel. The gesture concentration is, however, required to be digitalized for each point before the entire image can be processed as well as analyzed digitally. This means that the concentration of the gesture gathered at each point is required to be binned as also amplified.

There is a wide range of signals available due to the interaction between the specimen surface and the electron probe, and these signals power the scanning electron microscope. The varied signals include: the typical X-rays that are produced as an outcome of the excited inner shell electrons, the cathode luminescence wherein the electrons are excited in the scope of the observable light, the specimen current that passes through the sample, and the scattered back electrons that are inelastically as well as elastically dispersed out from the surface of the probe. The electrons ejected out from the surface of the target have low energy than the subsequent electrons and are evicted because of the interaction between the primary ray and the sample surface. These are detected readily and have the potential to form the image. The diameter of the focused probe limits the resolution of image produced.

The key distinctions between the scanning electron microscope and the transmission electron microscope can be succinctly summarized in relation to the modes of data collection that contribute to image formation. In the case of the transmission electron microscopy, the information is composed on the complete magnetic field view. A suitable lens system is employed to focus and build the magnified view, which is a function of the data collected, which is a function of the period. The information regarding the scanning electron microscope is gathered continuously as the probe scans across the specimen surface. For ensuring that the sample gesture noted at each specimen point is suitable, the rate of scan must be controlled appropriately. The whole period needed to scan an image is calculated by the product of the scanning rapidity for every pixel and the number of pixels to be scanned. Here it is worth noting that the distinction arising out due to the manner of data collection need not be overemphasized.

2.2 Basic principles

The basic principles that determine the basic behavior of electrons in the presence of a magnetic field and its interaction with matter have been delineated in the subsequent discussions.

2.2.1 Wave properties

The dual wave-particle nature exhibited by electrons makes it possible to focus as an electron ray. The de Broglie relationship expresses this wave-particle nature and is given by the following eq. (2.1)

$$\lambda = \frac{h}{mv} \tag{2.1}$$

where m is the mass associated with the particle, v is the particle velocity, and h is the Planck's constant. The electron energy, with the assumption that V is the accelerating voltage, is given by eq. (2.2):

$$eV = \frac{mv^2}{2} \tag{2.2}$$

where the charge of the electron is denoted by e. Therefore, $\lambda = h/(2meV)^{1/2}$ and this value is rather approximate. This is because the rest mass of the electron m_0 is very small in comparison to the relative mass m. As such, an alteration term is involved in the de-Broglie calculation and the corrected equation is given by eq. (2.3):

$$\lambda = \frac{h}{\sqrt{2m_0 eV \left(\frac{1 + eV}{2m_0 c^2} \right)}} \tag{2.3}$$

where the velocity of the light is given by c. The relative correction amounts to around 5 to 30%. The electronic wavelength at 100 kV is 0.00370 mm, which is two guidelines lesser than the interatomic space in the solid state.

Both the electrostatic as well as the magnetic fields aid in the deflection of the electrons. The deflected electrons can be focused by employing electrostatic and magnetic field geometry. The electrons in the region of the electron gun are affected by the creation of an electrostatic arena via the anode as well as the Wehnelt prejudiced cylinder. These typically result in an initial focus, which is referred to as the virtual electron basis. With this one exemption, the subsequent concentration in the electron microscope is based on the fundamental framework of the electromagnetic field and is achieved by employing an electromagnetic optic. The electromagnetic lenses are fortified with soft iron pole pieces. Unlike the case of the optical lenses that are made from glass, the electromagnetic lens has a variable focal length and can be regulated by making variations in the lens current.

An electron in the magnetic field is bent in a direction perpendicular to the level containing the magnetic field as well as the unique vector of the electron movement. An electron travelling off-axis will follow a helical path in case the magnetic field is uniform. Electrons with the same quantum of energy will be made to be focus at a point once they have traveled in conical directions from their source point in the uniform magnetic field.

The formation of image that takes place owing to the focusing electrons with the aid of electromagnetic lenses, differ in several aspects from that obtained using glass lenses. In the first instance, the formation of the image takes place perpendicular to the axis of the electromagnetic lenses for an object that is placed perpendicular to this axis. As such, focusing the objective lens by controlling the lens current gives rise to the spin of the image with respect to the same axis. Therefore, it can be concluded that the two images produced by two objects at dissimilar amplifications will be rotated about one another. Reversing the magnetic field can aid in compensating the rotation of images. This has become a common practice in designing the electromagnetic lenses and, as such, provisions are made in the lens system to transmit the current in the convolution of the optic coil.

There occurs a sudden alteration in the refractive index in the case of light optical microscope once the light is bounced as it enters the optic system. However, the refractive index remains constant within the glass lens. There occurs a continuous deflection of electrons in the case of the electromagnetic lens system. Moreover, the magnetic field generated by the optic pole, fragments endlessly through the lens system. The angle subtended by the electronic path with respect to the axis of the optic system is very small. Owing to this, the optic track over the electromagnetic field is long in comparison to the angular feast of the ray. Therefore, the transmission electron microscope has a small numerical aperture. On the other hand, in the case of the optical microscope, the numerical aperture of the objective might agree to the angle subtended by the light i.e., ranging from 45 to 90°. The numerical aperture of an electromagnetic lens is below 10^{-2}.

The physics associated with an electromagnetic lens reveals that modest geometric optics are just inadequate to describe the optic formation taking place in the case of the transmission electron microscope. The modest relationship between the amplification, the relative positions of the image and the object, and the focal length does not hold well in the case of the transmission electron microscope. This may be attributed to fact that the lens is slim in comparison to the entire optical pathway between the image and the object. The aforementioned approximation of the slim lens is likewise the factor for the inadequacy to describe the image formation in the case of high-powered objectives.

As such, ray diagrams are the means to exemplify the imagery means in electron microscope. Ray illustrations are 2-Dl representation of the electron path since the electrons are not rotated and the path of the electrons changes abruptly at the lens positions. It is considered that the thin lens approximation is still valid. In such a case of qualitative model, there is no quantitative calculation that is possible.

The extreme ray deviation in electron microscope is less than 1° and therefore Rayleigh criteria can be reduced to $\delta = 0.61\ \lambda/\mu sin\alpha$, that is, approximately $0.61\ \lambda/\alpha$, which is greater than $60\ \lambda$. The likely resolution of the transmission electron microscope must be in the range of 0.2 nm in case a wavelength of 100 kV is used, which is equivalent to 0.2 nm. However, with an advanced electrical energy, improved resolutions can be obtained.

For a depth of field, the light optical expression can be utilized as a gross approximation i.e., $d \sim \delta/\alpha$. With this approximation, the thin sample employed in the transmission electron microscope will be in range of 20–200 nm in thickness. Similarly, in the case of depth of focus, $D = M^2d$, for a magnification M of 10,000, it will be in terms of meters. As such, there will be no problem in recording the image of an object that is located about a span under the concentrating screen.

2.2.2 Resolution limits and aberrations associated with a lens system

The diffraction limit on the resolution is given by the Rayleigh criteria, $\delta_d = 0.61\ \lambda/\mu sin\alpha$. In the case of vacuum, $\mu = 1$, and at minor angles, $sin\alpha = \alpha$. The aforementioned expression changes to the following when the wavelength of the electron ray is expressed as rushing voltages i.e., $\delta_d = 0.61\ \lambda/\alpha = 0.75/[\alpha\sqrt{V(1 + 10^{-6}\ V)}]$. Therefore, the resolution can be improved by increasing the accelerating voltage for a given divergence angle. Accelerating voltages of up to 3 MV have been achieved; however, their commercialization has been restricted to 1 MV. At large accelerating voltages, radiation damages are suffered by the specimen. Most transmission electron microscopy is performed at 300–400 kV and this magnitude of accelerating voltage is close to causing radiation damage to the specimen being imaged. A point-to-point resolve in amorphous sample is available readily and is achievable routinely.

Similar to the light optic, an electron ray will be brought to the focus by employing an electromagnetic lens, given that the electron ray before passing through the electromagnetic lens is analogous with a certain span from the optical axis. On the other hand, an electron ray that is far from the optical axis will be fixated nearer to the electromagnetic lens. Under this situation, the level of the finest attention corresponds to the disc of minimum misperception. The dimensions of this disc of minimum misperception will, however, depend on the angular feast of the ray. This is known as sphere-shaped abnormality and the abnormality-dependent bound on the firmness is determined by the extent of the disc of minimum misperception, and is approximately given as $\delta_s \sim C_s\alpha^3$. The spherical aberration coefficient is represented by C_s. The deflection boundary on the firmness δ_d is in reverse relation to the angular aperture of the objective. On the other hand, the sphere-shaped abnormality limit δ_s is relational to the cube of the angular aperture.

It is revealed through various studies that for any given lens that has a fixed spherical aberration coefficient, there should be an optimum angular aperture, for which δ_s equals δ_d. Hence, the required angular aperture is considered to be a sensitive function of both the spherical aberration of the lens and the accelerating voltage. For an electromagnetic lens, the typical values for the coefficient of spherical aberration is less than 1 mm. A value close to 8×10^{-6} is obtained for the coefficient value of spherical aberration of 0.6 mm and for the wavelength corresponding to 100 kV of electrons. The design of electromagnetic lens has evolved over the decades and as such, multipole electrostatic spherical aberration correctors have been introduced into the design, and spherical aberration coefficient has been reduced to an arbitrary value.

Chromatic aberrations arise due to the lower deflection by the magnetic field in comparison to that experienced by the low-energy electrons. As such, the energy electrons are focused at a point that is far from the center of the lens, resulting into a disc of least confusion. There are multiple sources for chromatic aberration and the most common is one that arises because of the actively feast in ray energy. The relative energy feast is given by $\Delta E/E_0 = kT/eV$ in the case where the electrons are emitted thermally. Here, k is the Boltzmann's coefficient and e is the charge associated with electrons. The energy feast is close to 1.5×10^{-6} for T = 2,000 K and for 100 kV electron energy. There may also be some harm in the energy of the electrons owing to the inelastic scattering in the thin sample. In case the sample is thick, then an appreciable proportion of the electrons may get affected. Chromatic aberrations also may arise owing to the fluctuations in the current associated with the electromagnetic objective lens. Equation (2.4) delineates the relationship between the chromatic aberration limit on resolution:

$$\delta_c = C_c \frac{\Delta E}{E_0} \alpha \tag{2.4}$$

where the chromatic aberration coefficient is given by C_c and ΔE encompasses the uncertainties in both the quickening electrical energy associated with the objective lens. Similar to the spherical aberrations, the resolution limit increases linearly with α. The chromatic aberrations can be restricted by the temperature of the electron, given the adequacy in both the voltage of the electron cannon and the current associated with the lenses.

One of the extremely important factors is the axis of equilibrium of the electro-optical setup of the electron microscope. The performance of the instrument can be optimized by the precise positioning of the lens component within the microscope pillar. The objective lens is greatly influenced by the misalignment. The axis equilibrium of the lens system is highly subtle to disturbances linked at minor levels with the position, dimensions, geometry, and dielectric possessions of the specimen. The deposition of carbonaceous contamination onto the sample or onto the aperture of objective also results in minor disturbances to the alignment of the lens.

Difference in the focal span is one of the major reasons for the loss in axis equilibrium. The spinning electrons near the optical axis results into two principal focal locations on the optical axis. This in turn gives rise to two line foci at right angles. The condition is termed as astigmatism. The defect of astigmatism cannot be corrected owing to the intrinsic remaining unevenness in the lens system as well as due to the pole piece construction. This is also the purpose of the high sensitivity of astigmatism towards sample symmetry, contamination in instrument, and minor misalignment. However, correction can be achieved. Through the introduction of sets of correction coils, a complete twofold astigmatism correction is possible. The introduced corrections coils have their magnetic fields perpendicular to the optical axis as well as to the magnetic field of the two key lens coils. The adjustment to the rectification coil's current may be made in the midst of the functioning of the microscope and it balances any changes taking place owing to the magnetic asymmetry. The magnetic asymmetry may arise because of the buildup of the contamination or due to the specimen displacement while viewing it. On-line rectification is critical, particularly when operating with deep samples that possess high dielectric constant. Numerous arrangements for astigmator assemblage are possible, as for instance, the formation of an octad or octopole with four pairs of coils.

2.3 A comparative analysis between the scanning electron microscope and the transmission electron microscope

In the case of the transmission electron microscope, the image formation takes place by focusing the objective lens. The image is then enlarged by employing a series of additional imaging lenses, fluorescent screen, a CCD recording array, and a photographic emulsion. In the case of the scanning electron microscope, the image is shaped pixel to pixel through the collection of signals produced by the interface of the focused electron ray probe across the surface of the sample.

The numerical aperture, in the case of the transmission electron microscope, limits its depth and the resolution. However, owing to the small angular aperture of the electromagnetic lenses, the deepness of the arena of the transmission electron microscope exceeds their resolution by two orders of magnitude. The electron probe focusses the image in a scanning electron microscope and is analogous to the objective lens employed in a transmission electron microscope. The probe size is, however, restricted to the nanometer range, owing to the inelastic scattering processes occurring because of the interaction between the specimen and the electron probe. With an angular aperture of the probe lens of the size of 10^{-3}, the depth of field in the case of the scanning electron microscope is classically of the size of micrometers. The deepness of field that can be attained is therefore better than what can be achieved in the case of an optical microscope. A two-dimensional image is generated by the dual reason of light optical and transmission electron microscope. On the contrary, in the case of the scanning electron microscope, substantial in-focus data is produced that is associated with the three-dimensional topography of the sample under consideration.

The specimen, in the case of the transmission electron microscopy, is placed inside the magnetic field of the objective lens and hence fulfils electro-optical image requirements. The space that is available for the specimen is therefore very restrictive and as such restriction is on the adjacent sizes of the sample and additionally on the already extant boundaries on the thickness of the sample. The typical external sample dimension is restricted to 3 mm. On the other hand, in the case of the scanning electron microscope, the sample sits under the probe lens and is quite far away from the magnetic field. A sensible firmness is accessible even when there is a parting of 50 mm between the lens and the sample. Moreover, there are no boundaries on the adjacent sizes of the sample. Only the constraint on the lateral dimensions comes from the geometry of the specimen compartment. Most specimens have adjacent sizes ranging from 20–30 mm; however, specimen chambers have been designed to accommodate specimens that are 10 cm or more in size. Common examples include that of unsuccessful manufacturing mechanisms and solid-state devices.

There are three factors that determine the vacuum requirements in the case of electron microscopes. The first factor is associated with the need to evade the high-energy electrons from scattering owing to the presence of remaining gas in the microscope pillar. The second factor concerns the chemical as well as thermal steadiness of the electron cannon during the operation of the microscope. The last factor is concerned with the need to minimize or remove the ray-induced adulteration of the specimen. The necessity on vacuum is least stringent owing to the fact that a void of 10^{-5} Torr is fairly adequate to safeguard the negligible smattering of the electrons that takes because of the formation of the magnetic field. The second factor is, however, very critical. Erosion of the tungsten electrode takes place through oxidation when operating under a vacuum of 10^{-5} Torr. The other alternative sources i.e., lanthanum hexaboride (LaB_6) crystal as well as the field emission source requires a superior void, of the order of 10^{-7} Torr as well as 10^{-10} Torr, respectively.

The third factor is equally critical. The contamination of the specimen results due to the inelastic interaction between the high-energy electrons incident on the surface and the contaminant gases absorbed on the surface. An adherent, carbonaceous, and amorphous layer is formed on the specimen surface as a result of the pyrolysis and polymerization of the hydrocarbons that forms on the specimen surface. The deposited carbonaceous layer may lead to obscuring of morphological details, if present for an extended period of time over the surface.

Cryogenic cooling is one of the suitable means to avoid contamination of the specimen surface. The procedure is generally adopted in the case of transmission electron microscopy wherein the condensable contaminants are trapped. Cryogenic trap is more effective in the case of scanning electron microscopy because of the large specimen dimensions used. However, the rate of contamination is exacerbated because of the high-energy focused current in the electron probe. Plasma etching is one of the solutions to ensure the specimen is free from contamination. This is done in the presence of oxygen and argon gas mixture that oxidizes the carbon deposits on the specimen surface. The plasma etching process can be carried out in the best possible vacuum chamber to yield better results.

The images produced using scanning electron microscopy may be distorted owing to the electrical instability associated with the systems used in scanning. Major outcomes may be affected in a significant manner and this may include any one of the following: alterations in the operative magnification in the x as well as y scan directions, drift from the image, which is the need for amplification, on the span from the optic axis, the likely shear alteration of the image, particularly in x-direction, barreling, in case the center area is amplified, and the pin-cushion influence, in case the enlargement is for the peripheral region. In a majority of the cases, the imaging flaws produced are an outcome of the pixel to pixel information gathering procedure and may also be because of the alterations in the scanning raster of the probe's x-y coordinates. Another major source of instability may be the electrostatic charging of the sample i.e., the sample being an electrical insulator. This may be prohibited by a conductive covering or by employing very small ray currents as well as voltages.

2.4 Sample preparation

Sample preparation is one of important and tedious tasks in the case of transmission electron microscopy. However, there are various techniques to prepare a good sample for transmission electron microscopy and yield appropriate results. The information obtained using transmission electron microscopy can be unique in the case of good specimen preparation. It is frustrating to expect better information from a sample that has been prepared poorly.

Successful and efficient transmission electron microscopy is greatly dependent on three diverse skills: preparation of good sample, acquiring of appropriate and useful info, and having the skills to understand the gathered info. Specimen preparation has been made easy with the wide possibility of trials and the existence of decent marketable kits. With the availability of good facilities, there are no barriers to prepare good specimens – thin layer samples from a given engineering material – and no explanation on the deteriorating period on inspection of poorly prepared samples. Good samples for transmission electron microscopy need to be less than 100 nm in thickness, and in the case of lattice imaging in high firmness or sub-micrometer micro level analysis by electron energy loss spectroscopy (EELS), the sample thickness must be less than 20 nm. Preparation of specimens of such thickness is possible, however, only if adequate tools are available.

In the ensuing discussion, there is a brief on the making of thin films from bulk engineering materials. One of the important tasks that can be accomplished using transmission electron microscopy is the characterization of the powder samples. The trick lies in the preparation of samples that reflect the composition, particle size, and shape without the introduction of artifacts that are associated with fragmentation or agglomeration. The preparation usually involves preparing a steady dispersal in a liquid medium, and is accomplished using surface-active essences. A droplet of the prepared dispersal is typically used on a smooth carbon layer that is kept on a microscopic mesh. However, as the elements gather at the meniscus, the preparation procedure is seldom employed. The collection at the meniscus leaves irregular aggregates that are difficult to be interpreted. The more suitable technique lies in spraying the dispersal over the carbon-covered mesh with the aid of a nebulizer. The size of the sprayed droplets is only a few micrometers in diameter and there will no more than two particles in case a sufficiently diluted dispersion is prepared. As such, the danger arising due to aggregation is avoided and the size distribution obtained will be characteristic of the original dispersion.

A microtome is employed to section the soft tissues and polymeric samples. The microtome prepares a sequence of thin slices from the solid prepared sample, which is similar to that of sliced bread. An examination of the serially ordered sequence of slices results in building up of an image of the 3-D assembly of the sample. Glass knives are also employed with the approach of the sample stub being to regulate the thermal growth of the mounting rod. A diamond knife is also used with a piezoelectric regulator for the preparation of the slim slices.

2.4.1 Mechanical thinning

The specimen making for transmission electron microscopy starts with the thin-film sample sliced from a bulk specimen. The diameter of the thin film is typically 3 mm and its thinness is in numerous hundreds of micrometers. The thin-film disc-shaped specimen might be stamped out from a ductile metallic piece, cut out from a bar, trepanned from a brittle ceramic, or may be taken out by machining from a greater segment. In all the aforementioned cases of disc preparation, it is important to choose the appropriate symmetrical axis and the center of the prepared disc as regards the coordinates of the bulk sample. This is because the choice of the location makes sample making easier and hence watching through the microscope. In this point of sample preparation, it is critical to minimize the mechanical harm to the material and prevent the flatness and smoothness of the surface that is cut out.

The subsequent step is the reduction in thinness of the disc specimen. Grinding and polishing can be employed to accomplish the thickness reduction task. The choice of the grinding media is determined by the stiffness, toughness, and the hardness of the material. The following techniques can be used to accomplish the mechanical thinning stage: (1) a polished and parallel side disc is equipped and is then made thin from single or both sides as per the requirements of planar geometry. The specimen is fixed to the flat and polished specimen using a crystalline wax. The wax is melted out in both scenarios: when the sample is affixed and when it is flipped onto the baseplate. The grit size must be decreased by reducing the thickness. The thickness of the sample should then be less than 100 micrometer. The greatest challenge at this stage lies in relieving the sample of the residual stresses that may otherwise result in clasping off the specimen, once it is detached from the baseplate. (2) The disc is protected on an optically flat surface once the sample is reduced to 100 micrometer or less and then dimpled, wherein a good grinding media eliminates the material from the central area of the disc. The process is analogous to a lapping process, wherein the particles of the grinding media are expatriated in the region of the contact shear. There is a wide range of dimplers that are available to successfully thin the specimen of hard as well as brittle materials. Dimpling is one of the techniques that takes care of the distortions caused by the residual stresses. (3) Preparation of wedge may be preferred over a dimple in the case of some samples. This is the case for investigation of the interface of the microstructure. As such, the extreme slim zone from the border area can only be found when the border is straight and parallel to the edge of the wedge. It is a preferred option to segment the sandwich perpendicularly in the case of a slim layer and create multiple coat sandwiches, consisting of dissimilar stages. The plane of the sandwich is mounted vertical to the axis of the wedge. Every coat is divided as a wedge and the associated morphological characterization of the interface of each layer is then calculated in a solo slim layer specimen. A rigid jig is employed to prepare the wedge, wherein the wedge angle is already chosen and is usually kept at lesser than 10°. (4) The crosswise segments associated with the microelectronic instrument can be ready by either of the dual standardized methods employed widely in industry. In one of the methods, the instrument to be partitioned is diced into squares. The diced portions are then pasted to each other and a block of 3 mm thickness is formed. Subsequently, a rod is trepanned from the square slab, with the rod axis being in plane with the one-on-one layers. The trepanned rod is pasted into a 3 mm diameter metal tube, which is made of brass or copper. The tube is then divided and mechanically thinned to obtain a series of sections that are 100 μm thin. The sections are then dimpled and subsequently ion-milled.

2.4.2 Electrochemical thinning

In the case of the mechanical thinning process, the mechanical damage that takes place at the sub-surface cannot be avoided. Chemical dissolution is typically preferred over the mechanical thinning process in case the material is a metallic conductor. This is attained electrochemically. The methods established for electrochemical thinning are built on electropolishing results, though there is a wide difference in the conditions for electropolishing bulk components and the sample to be prepared for transmission electron microscopy. This is because in the case of an electrically conducting specimen, the area to be thinned is very small, and current with higher densities are more often used than that used for electropolishing of bulk material. A jet of polishing solution can aid in minimizing the problem of heat dissipation that is generated as a result of chemical attack at high current densities.

The current densities required and the temperature of the electropolishing solution are the critical factors in the successful preparation of the sample by jet polishing. These can be obtained from the manuals supplied by the manufacturers of jet-polishing units. The electrochemical thinning process is accomplished as soon as there appears a central hole on the disc sample.

The electrochemical thinning process is completed just after the initial pit appears at the center of the disc specimen. The formation of the pit is initially spotted by the eye and further typically via employing an optical laser gesture through which current is switched off and hence the jet, as soon as the optical laser passes through the pit. The specimen is then ready to be immersed into the microscope once it has been rinsed and dried completely. The regions that surround the central hole should be transparent to the electronic ray. If the area available for investigation is too small, this signifies that the jet polishing procedure was permissible to endure even after the development of the initial pit. The continual jet polishing procedure results in a rapid attack at the edge of the hole and hence the turning of pit. In case the polishing solution is exhausted, a roughened, etched appearance of the specimen surface is achieved. This may also arise due to contamination, overheating, or insufficient current density.

2.4.3 Ion milling

Raymnd Cataing in France was the pioneer of the ion milling process for thin layer samples from ductile metals and alloys. A ray of exited inert gas was employed to splutter away the surface of the thin-film aluminum alloy specimen. The ion milling process was later on superseded by chemical thinning methods. However, with the recent advancements, sophisticated ion milling machines have been developed, and the ion milling process is now one of the most preferred methods for removal of the final surface layers from a specimen for transmission electron microscopy.

There are numerous advantages of the ion milling process. It is a gaseous phase process and as such the contamination is easier to be removed from the surface. Ion milling process is more normally appropriate to ceramics and semiconductors while the electrochemical process is restricted to only metallic conductors. A suitable choice of the milling parameters can aid in minimizing the sub-surface radiation damage. The critical parameters involved are the ion energy as well as incident angle associated with the ion ray. The ion milling process also eliminates the impurity in films and there is no change in the surface composition of the specimen. This is because the ion milling operation is performed at a temperature well below the temperature where the dispersal happens in the sample. The availability of sophisticated ion milling equipment ensures that the thinned region for the specimen under consideration might get local, superior to 1 μm, which is one of the main considerations to study slim layer microelectronic as well as other associated optronic instruments. The exactness provided by ion milling in the selection of the localized area for electron microscopy is not possible while employing the chemical approaches.

While there are advantages, there are obvious disadvantages associated with the ion milling process. Splattering is a momentum transmission procedure and its frequency is extreme when the ion ray is perpendicular to the specimen surface. The atomic mass of ions associated with sputtering is near to the specimen surface. As such, the ion ray at right angles may lead to sub-surface radiation damage and also surface irregularities. The rate of sputtering can be enhanced by raising the energy of the incoming ion ray. However, doing so comes at a rate of suppressing a large number of splattering ions within the sub-surface area. This in turn results in radiation damage of the subsurface layer. Hence, to take care of this aspect, the energy associated with the ion ray is inadequate to a few kilovolts. At these limited drives, the deepness of the ion ray dose is inadequate, which allows for most of the sputtering ions to outflow to the surface by the process of dispersal and therefore avoid sub-surface harm.

The incident angle of the ion ray is limited to 15° at an energy of nearly 5 kV. The splattering rate is limited to 50 μm h^{-1}. The chosen sample for transmission electron microscopy is almost thin, ranging from 20–50 μm, thinned mechanically and using the ion milling process. The disc sample is rotated in the ion milling process, which ensures that the thinning takes place uniformly. Initially, the thinning process is allowed to take place on both sides of the specimen and the angle of incident ion ray is kept at 18°. Then, over subsequent steps of the thinning procedure, the incident angle of the ion ray is reduced so as to minimize the surface roughness. The geometry of the ion ray dictates the minimum incident angle of ion ray that will produce optically planar surface finish. At glancing angles, there may be chances that the ion ray may splutter material from the sample mounting assemblage as well, thereby resulting in contamination of the specimen. This is the main reason that the minimum sputtering angle ranges from 2° to 6°.

The milling process is completed as soon as the first hole is shaped in the specimen. To ensure accuracy in the ion milling process, a light transmission detector monitors the area that is to be thinned. Just after the required region is perforated, the milling process is terminated. Unfortunately, the process may not be feasible if one of the films in the specimen is transparent.

2.4.4 Carbon and sputter coating

The specimen that is nonconductive to electricity develops electrostatic charge when struck by the electron ray. In many cases where specimens are being investigated, charging is not a problem since the charge can be limited by the small size of the sample. However, if required, the sample can be layered using an electrical conducting layer. The most favored conducting material for coating is carbon because it has a low atomic number and gets deposited uniformly to form an amorphous slim layer. Any microstructure that emerges as a result of coating has an extremely little contrast on a nanoscale.

The passing of large electric current through the point of interaction between the dual carbon bars may aid in the evaporation of the carbon coating onto the surface. The specimen may also be sputter coated by the bombardment of the specimen surface by inert gas ion. There are samples such as ceramics that may require coatings on both sides. The deposited coating is of the nanometer dimension, ranging from 5–10 nm, and is barely visible in the recorded microscopic image.

2.4.5 Replica methods

Taking a replica from the surface of the specimen is an alternative to the preparation of a thin slice from the specimen. This may not be necessary in case the scanning electron microscopy provides images at sufficient resolutions. However, in the case of the transmission electron microscope, there may be several reasons why the replica method is desirable. One of the reasons may be the nondestructive examination. There may be circumstances, as in failure analysis, or for logical reasons that there may arise the reluctance to section the component. Therefore, taking a replica is more desirable wherein it can be taken far away from the laboratory. Forensic inquiries at crime scenes and also from accident sites are situations of this kind. In such cases, it is critically required to present the evidence in courts without causing any damage to the collected evidences. Complex samples is another reason that justifies the desirability of sectioning the specimen and then analyze it through transmission electron microscopy. This is convenient when analysis is sought for the specific phase present on the surface of the specimen. Examples of this case can be observed in forensic domains such as the gunshots that are recovered from the skin, or the particles associated with paint pigments in accidental cases. In many cases, corrosion products can be observed under isolation. This is because the corrosive product will not affect the composition and structure of the bulk material significantly and hence can be observed through the replica method. Other instance that suits the narrative of replica methods is the examination of detailed stages from a polyphase material. The specific phases can be isolated from the bulk material through suitable chemical etchants that ensure that the original phase distribution in the original specimen is retained. Transmission electron microscopy can then be used to analyze the chemistry, morphology, and crystallography of the extracted phases. The remaining bulk material does not interfere with the examination prospects. The instance for this case includes the carbide precipitation of steels wherein the morphology, crystal structure, and the composition can be revealed for the carbide phases. This analysis is observed to be obscured in case it has been carried out through X-ray excitation and diffraction methods. One of the last but not the least cases wherein the replica method justifies itself is the correlation of the microstructures by employing alternate imaging methods. That means, there may be situations when it is required to compare the observations made on replica through transmission electron microscope and that made using the scanning electron microscope for the same surface. Instance of such situations include the surface markings due to mechanical fatigue, wherein a combination of the two aforementioned imaging techniques may be advantageous. Thus, the scanning electron microscopy can be employed to observe ductile failures whereas the transmission electron microscopy can be used to decipher the nonmetallic inclusions associated with the nucleation of the dimples.

The usual procedure includes taking a negative replica on a pliable and solvable plastic. The employed plastic might be cast in a position and hence allowed to become hard. In another instance, a plastic sheet that has been softened may be used. The softened plastic sheet may be hard-pressed onto the surface earlier, allowing the evaporation of the solvent. In many cases, the ultrasonic procedure may be used to remove the loose contamination. In many other cases, the contamination may be the subject of observation as is the instance with the remains associated with gunshots.

The plastic replica may be skinned from the surface, on hardening, and then weighty metal can be employed to cast a shadow. The heavy metal can be a gold-palladium alloy and the shadowing enhances the last contrast in the electron microscope. The shadowed metal is carefully chosen for maximum scattering power as well as for minimum particle size. The size of the particulate cluster is about 3 nanometers. A carbon film 100–200 nanometers in thinness is placed on the plastic imitation. The plastic is liquefied by employing appropriate organic solvents. The carbon layer ensures that any particle that is possibly removed by the plastic is retained suitably on the specimen surface. Moreover, the particles that may have been removed, owing to shadowing through heavy metal, are also retained by employing a carbon film. The carbon imitation is then washed thoroughly and is composed on a fine-mesh grid of copper. This is finally viewed in the transmission electron microscope.

The negative plastic replica may not at all be required. As for instance, there may be an alloy sample wherein the surface is polished to reveal the particles of the second phase that can be layered directly with the carbon coat placed straight on the surface as well as that adheres sturdily with the particles. The carbon extraction replica is revealed through the etching process on the matrix that is distributed in its original configuration.

2.5 The origin of contrast

The electron ray interrelates with the sample elastically as well as inelastically. However, the contrast observed through the observation in transmission electron microscopy is revealed through the elastic interactions. The inelastic interactions, on the other hand, carry the information associated with the chemical composition of the sample. The contrast arises owing to the three distinct image formation processes: phase contrast, diffraction contrast, and mass-thickness contrast. In the case of the amorphous sample that is with the specimen containing the glassy microstructure and in the absence of the long-range crystalline order, there arises an envelope of transmitted intensity through elastic scattering. This envelope varies with the scattering angle, in accordance with the $\cos^2\theta$ law. The intensity of the scattered-out electron ray depends to a great extent on the energy associated with the electron ray, the thinness of the sample, and the density of the sample. Therefore, it is the mass-thickness that reflects on the image contrast. The aperture will hinder the greatest of the dispersed electrons in case the aperture is positioned in the electro-optical pillar and herein, the image will be dominated by the electrons transmitted directly. Mass-thickness image formation process is dominated in the case of biological samples observed by using transmission electron microscopy.

In the case of crystalline samples, the scattering of electrons takes place elastically in accordance with the Bragg's law. The scattering process generates deflected rays at distinct angles $2\theta_{hkl}$ to the ray conveyed directly. This corresponds to the crystal plane for which the Miller-indices is equivalent to *hkl* and the Bragg's law is satisfied. An aperture might be put into the optical pillar at the plane of the image under the sample, also allowing the straight conveyed ray to enter the imaging setup and therefore form a bright-field image. There may be chances that the diffracted ray may be accepted in order to obtain the dark-field image. In either case, the image contrast is resolute by the existence of flaws associated with the crystal lattices, affecting the locally diffracted intensity. The aforementioned imaging system or means is termed as deflection contrast. The mass thinness as well as the diffraction contrast essentially create the magnified shades of microstructural characteristics. This can be compared with the extended shadow of a tree on the grass wherein the image of the branches as well as leaves are reflected as a two-dimensional projection, depending on the angle of the sun.

An aperture with larger diameter can be inserted in case there is adequate resolving power associated with the microscope. A large diameter aperture will allow several diffracted rays to be admitted into the imaging system. A lattice image is produced, reflecting on the periodicity in the crystal structure. This effect is termed as phase contrast. In the case of phase contrast, elastically scattered electrons are used to create the image of the crystal structure. The lattice image is incomplete because of the noninclusion of all the diffracted rays.

2.5.1 Mass-thickness contrast

The possibility with which an electron may be elastically dispersed out depends on the factor termed as atomic scattering factor. The atomic scattering issue surges with the atomic number of atoms as well as their number in the path of the incidence electron ray. Therefore, it is the total thickness of the film that dictates the chance of an electron being dispersed out. Mass-thickness contrast therefore reflects the combinatorial dependence on the density as well as the thickness of the specimen. In the domain of life sciences, the mass-thickness contrast dominates in comparison to the other two contrasts. Heavy-metal and tissue staining procedure is adopted to enhance the contrast associated with the soft tissues as well as the biological samples. Mass-contrast dominates in the cases of noncrystalline structures that lie in the domain of engineering and natural sciences. Also, in the case of replica studies, the mass-contrast dominates, owing to the differences in the thinness of metal shadow that gets placed onto the replica.

2.5.2 Diffraction contrast

The amplitude of the scattered incidence ray into the deflected ray dictates the contrast in the microscope in the case of a perfect crystal. The amplitude can be obtained as a summation of the deflected amplitudes from the unit cell lying along the pathway of the deflected ray. Equation (2.5) provides a relationship between the phase difference ϕ that has been dispersed by the unit cell located at site vector r with respect to the origin of the pillar of material that is answerable for producing the contrast:

$$\phi = 2\pi(g \cdot r) \tag{2.5}$$

Equation (2.6) on the other hand provides for the determination of the amplitude scattered by the unit cell:

$$Ae^{i\phi} = A \exp[-2\pi i(g \cdot r)] \tag{2.6}$$

The lattice parameter a is the defining distance between the unit cells in the pillar of crystal. $g \cdot r = n$ holds true in the case of the Bragg diffraction, which is the case for in-phase scattering. Every unit cell in the entire crystal structure scatters on its own, owing to the smaller wavelength of the electron ray in comparison to the aforementioned lattice parameter a. The amplitude being scattered by the single unit cell is the summation of the atomic scattering factors.

Any reduction in the amplitude of the ray transmitted directly can be ignored in case the amplitude dispersed is minor in comparison to the incoming amplitude. As such, every unit cell in the column disperses the amplitude of the similar order. The entire amplitude dispersed by the pillar comprising of unit cells is a lined role of n at the Bragg's position, and can be depicted as follows using eq. (2.7):

$$A_n = \sum_n F_n \exp[-2\pi i(g \cdot r)] \tag{2.7}$$

In case the numeral of the unit cells in a pillar is adequately large, the summation can be replaced with the essential, and for the sake of the convenience, the thinness of the slim coat can be measured from the mid thickness. As a result of which, eq. (2.7) can be rewritten as follows:

$$A_t = \frac{F}{a} \int_{-t/2}^{t/2} \exp[-2\pi i(g \cdot r)] \tag{2.8}$$

The scattering vector g is replaced with $g + s$ in case the angle of incidence deviates from the Bragg state. The angular deviation in the above replacement is reflected by s and the deviation is measured from the Bragg position in the reciprocal space. The location of the dispersing element that is initially at location r in the pillar shifts to $r + R$ in case there is distortion in the lattice, because of the lattice defect present in the crystal. The phase angle associated with the amplitude scattered can then be given by the following formulation represented by eq. (2.9):

$$\phi = 2\pi(g + s) \cdot (r + R) \tag{2.9}$$

The expansion of brackets in eq. (2.9) results in the following simplified eq. (2.10):

$$\phi = 2\pi(g \cdot r + g \cdot R + s \cdot r + s \cdot R) \tag{2.10}$$

The term $g \cdot r = n$ is a numeral term and therefore has no consequence on the stage of the largeness that is dispersed. The value $s.R$ can be neglected as it is the product of two small vectors. The residual values $g \cdot R$ and $s \cdot r$ are added, and signify the stage change in the largeness that is dispersed at place r into the deflected ray g. This may arise either due to the deviations from the precise Bragg conditions, $s \cdot r$, or may be due to the distortion of the crystal lattice, which signifies the strains in the lattice, owing to the existence of lattice flaws, $g \cdot R$. It is worth noting here that in the microstructural characteristics imaged by the deflection contrast, the user is viewing the summation of the contrast effects produced due to the deviations in the reciprocal space and the displacements taking place in the real space.

For a transparent interpretation of the diffraction contrast, it is required two identify the two aforementioned defects i.e., their reason for occurrence, the functioning reflections, the g vectors, and the reasons for displacements, both in the reciprocal and in the real space. This is a tough procedure and requires information with regard to the thinness of the specimen, in addition to the knowledge with regard to the bright-field and dark-field images that have been obtained through different g reflections. It is not at all required to have a complete understanding on the aspects associated with the diffraction analysis. While an understanding on the image analysis may be restricted to the generic recognition as pertaining to the lattice flaws, stacking fault, and point defects, a whole measurable study of a flaw can be avoided, as for instance the sign and the magnitude associated with the Burgers vector or the value associated with the stacking fault vector.

2.5.3 Phase contrast

The transmission electron microscopic instrument works either as a diffraction camera or as a microscope. The angle of receiving for the ray is limited by an objective aperture that has been sighted in the backward focal plane of the objective lens in case the transmission electron microscope is operated as microscope and the imaging is accomplished either by diffraction contrast or by mass-thickness mode. The acceptance angle is lesser in comparison to the Bragg angle associated with any of the reasonably dispersed rays. The image produced is then a bright-filed, shadowed projected image of the electron that comes from the objective aperture. The intensity variation in the obtained image plane suggests disparities in the plane of the image associated with the electron ray current that passes from the microscope column and then crossing the dissimilar areas in the field of sight on the thin-film specimen. The diffracted ray is accepted into the system when the incidence ray is slanted by a Bragg angle and then the straight conveyed ray passes through and downward the optic axis of the microscope. The bright field image is now substituted with the dark filed image once the objective aperture cuts out the diffracted as well as the transmitted rays.

An interference pattern will result in case the angle of acceptance is more than twice the Bragg angle i.e., the objective aperture is detached and substituted with the aperture that is sufficiently big to admit the diffracted ray as well as the transmitted ray into the imaging system. Several characteristic parameters are required to be known in case an interpretation on the interference pattern is required. These parameters comprise the values associated with the chromatic and sphere-shaped abnormality coefficients, the precise know-how on the focusing plane of the image, and the consistency and the energy feast in the incoming electron ray that is identified by the source of electron.

It is important that the stage contrast in the case of the transmission electron microscope agrees with the interference pattern. Imagine that there is an object that is signified by a number of point sources and $f(x, y)$ represents the total electron wave spreading in the plane of the object. The function corresponding to this aforementioned function that signifies the amplitude and phase is represented by $g(x, y)$ in the level of the image to the object under consideration. Each point in the plane of the image will consist of aids from all the rays that have been conveyed and therefore the following eq. (2.11) will hold:

$$g(r) = \int f(r')h(r - r')dr' = f(r) \times h(r - r') \qquad (2.11)$$

As can be observed, a more suitable radial coordinate r has been employed in the place of using the Cartesian coordinates. The influence of the electron wave impacting each of the distinct objects is represented by $h(r)$ and is named as a point-spread job or the impulse response function. Convolution of $f(r)$ with $h(r)$ is represented by $g(r)$. The electron ray density distribution in the specimen therefore modifies the electron wave job of the incidence ray. This is also convoluted with a function that provides a description of the response associated with the microscopic column comprising the electron ray source, the aperture size, the electromagnetic lens system, and the lens aberrations.

The reciprocal space can be moved conveniently wherein $g(r)$ can be represented by the Fourier transformation as follows:

$$g(r) = \sum_{u} G(u) \exp(2\pi i u \cdot r) \qquad (2.12)$$

where the reciprocal lattice vector is represented by u. The Fourier transformation of $f(r)$ can be represented as $F(u)$ as well as $h(r)$ as $H(u)$ and the following relationship hold true:

$$G(u) = H(u)F(u) \qquad (2.13)$$

where the contrast transfer function is represented by $H(u)$ and is typical of the microscope. The contrast transfer function is a nondimensional parameter and can be designed as a function of r^{-1}, which represents the role of the 3-D frequency attainment in the plane of the image. The contrast transfer function describes the influence on the stage shift of all the microscopic parameters. As such, three major contributions of the contrast transfer function can be highlighted: an aperture function that describes the cutoff boundary associated with the 3-D incidences, and is determined by the radius of the aperture, a cover function, reflective of the restraining associated with the 3-D frequencies and that is either due to chromatic abnormality or due to the uncertainties associated with the objective lens current, an abnormality function that bounds the 3-D frequency accessible for imaging and is conquered by the coefficient of sphere-shaped abnormality associated with the electromagnetic lens. The abnormality function can be calculated using the following eq. (2.14):

$$B(u) = \exp[ix(u)] \tag{2.14}$$

where

$$x(u) = \pi \Delta f \lambda u^2 + \frac{1}{2} \pi C_s \lambda^3 u^4 \tag{2.15}$$

In the above equation eq. (2.15), the under-focus of the objective lens can be described using Δf and is defined as the distance between the focal plane of lens as well as the objective plane, the wavelength is denoted by λ, the coefficient for spherical aberration is denoted by C_s and is characteristic of the objective lens.

The contrast transfer function can be shifted to improve the stage contrast for the precise and chosen crystallographic planes. This can be accomplished by controlling the defocus of the objective lens. The best performance can be achieved from the microscope only by a careful calculation of the contrast transfer function and through the careful analysis of the information being sought just before the inception of the microscope imaging session.

No information is passed to the image when the contrast transfer function is associated with a zero value; the contrast transfer function corresponding to the best performance of the objective lens will encompass certain oscillations about the zero values. The first contrast transfer function crossover will move to larger values of nm^{-1} in case the objective lens is under-focused. In such cases, the spherical aberrations associated with the objective lens can be compensated partially. Scherzer defocus is the terminology that has been provided to signify the position of the optimum compensation. Equation (2.16) aids in the determination of the Scherzer defocus:

$$\Delta f_{sch} = -1.2\sqrt{(C_s\lambda)} \tag{2.16}$$

The resolution limit of the microscope is determined by Scherzer defocus and is termed as point resolution. This agrees to the least defocus worth, at which all the rays that are below the first contrast transfer function cusp will possess a persistent stage. The info associated with the crystallography aspect will still be available in case the resolve bound of the microscope is below the Scherzer defocus. However, in order to interpret the complete information, entire computer simulations of the image are required. Below such circumstances, the information transferred from the microscope to the image is damped severely. As a result, the availability of the contrast from the crystallographic planes having minor d-spacing is inadequate. Therefore, a second resolution boundary, known as information limit, is set. The major reason behind the ready availability of the lattice images from the crystallographic planes with *d*-spacing below the point resolution is the difference among the point firmness and the info boundary. Point resolution is one of the major reasons for the working of microscopists on the cellular samples and biological tissues that exhibit noncrystallinity. However, for the materials community, the importance is the information limit, as this describes the least interplanar spaces that can aid in resolving the lattice image.

Spherical aberration corrections have been met with the introduction of hexapole electrostatic stigmators and since their introduction, the microscopist community was impacted significantly. The correction system has the potential ability to diminish the worth of sphere-shaped abnormality coefficient and therefore the point firmness as well as the resolution limit coincides. Resolutions as low as 0.07 nm have been achieved using the monochromatic electron source that has aided in the reduction of the chromatic aberration.

It is also conceivable to extract the info associated with phase from the lattice image through the added advantage of the resolution limit. As such, the information that is derived from the lattice structure is not restricted to the 2-D spreading of intensity in the *x-y* plane. When this addition was nonexistent, the series of thorough images and their analysis aid in the achievement of the aforementioned objective. Lattice imaging has advanced the technology that has brought the electron microscopists away from the shadow images that were based on the diffraction contrast and mass-thickness. It is worth realizing the importance of computer interpretation for the lattice contrast for appropriate image analysis.

2.6 Lattice imaging at high resolution

The available transmission electron microscopes have the potential to produce the phase contrast imaging of the crystal lattices as well as the associated defects. The main mode of operation as well as the related features with phase contrast imaging have already been described; in particular, the sensitive dependency on the various physical constraints such as that defined using contrast transfer role of the microscope. The following discussion is made on the association between the observed lattice images as well as the crystal structure.

2.6.1 Lattice image and contrast transfer function

The stage changes that are presented by the imaging setup of the microscope are summarized by the contrast transfer role. The phase shifts take place in the wave role of the electrons that form the image in the plane of the inspection zone and the imaging system of the microscope, which consists of the following: a CCD, a photographic recording emulsion, and a fluorescent screen. *sinx* characterizes the phase shift associated with the imaging setup of the microscope. The contrast transfer function provides *sinx* as a function of the spatial frequency or the wave number. This is proportionate to the scattering angle subtends by the rate at the aperture of the objective lens, and is measure in reciprocal space as 1/d. The convolution of the amplitude as well as the phase with the contrast transfer function provides the closing phase and largeness for each of the associated spatial frequencies. The phase shifts and the amplitudes are designed at the departure plane of the slim layer object and are determined by employing the dynamical theory of electron diffraction. These are determined before the electron ray is conveyed over the imaging setup of the microscope and till it hinders the imaging plane. The concentration is obtained at each point through the multiplication of the integrated image largeness by its multifaceted conjugate.

The issues that make up the contrast transfer function include those that are related both with the objective lens and the electron ray source. Coherence and energy spread are the source parameters. Poor coherence of the source results in large dimension of the LaB_6 crystal electron emitter. On the other hand, excellent coherence is exhibited by the field emission cannon that possesses very small diameter, which can be approximated to a point source. Moreover, the field emission gun has a very minor energy spread. The outdated tungsten filament is smaller with respect to both the emitting diameter as well as the energy spread. The source diameter of a traditional tungsten filament is very big and its working temperature is too high. Angular spread is another additional parameter that can be determined by the conditions related to the emitting ray and condenser system.

The limitations associated with the objective lens include the sphere-shaped abnormality coefficient as well as the current steadiness in the lens. The second- or third-order defects of the lens are additional correction factors. A second-order defect, for instance, *Astigmatism*, is removed readily and the third-order defect such as *coma* is now removable in the present sophisticated systems. Spherical aberration coefficient is the major parameter that limits the point of resolution of the microscope. However, the availability of the electrostatic spherical aberration correctors can aid in the removal of the present limitations.

2.6.2 Computer simulation of lattice images

The lattice image in phase contrast is noted as near as possible to the Scherzer focus. This is the under-focus that provides to maximize the 3-D rate for which the electron microscope can image the system and introduce swaying into the sign of stage move. The periodicity associated with the image of the lattice will be retained only when the images are taken serially and with the focus changing incrementally. The intensity of the image at each position will only change during such sequential operation. The change in intensity may include any one or the combination of the following aspects:

contrast setback at the defocus standards that are under the Scherzer focus, changes occurring apparently in the specific crystallographic direction, and copying of the distinguishing cyclicities associated with the lattice. The common observation includes the rows of bright or dark maxima in any of the way of the lattice and substituted by alike row characteristics in a dissimilar way of the lattice. Interpretations of the aforementioned observations are, however, not straightforward.

The assigning of specific features to the significant characteristics in the lattice image from the sample crystallography depends on the knowledge related to both the crystallographic structure as well as the observed diffraction patterns. Any scientist or microscopist can be made aware of the dangers associated with recognition of the features only if some practical experience is possessed with regard to the sensitivity of the lattice image contrast to the thickness of sample as well as to the defocus of the objective lens. Any discrepancies associated with the imaging of the artifacts can also be ascribed.

Computer simulations can be employed to evaluate the image and is the only acceptable scientific procedure. With the advancements, several software packages are now available to deal with the intricacies associated with the evaluation of the obtained image. Three conditions must, however, be satisfied for a fruitful study of the image contrast: (1) precise standardization of the limits associated with the microscopic instrument – this includes the determination of ray divergence and the chromatic and spherical aberrations. (2) The sample available must be free from contamination, sufficiently thin, and should not be thicker than the annihilation thickness that is required for the desired imaging reflections. (3) Accuracy in the placement of the optic axis of the microscope – it must be maintained along a significant axis of the specimen and to that of the angular aperture of the objective lens that can receive as many reflections as possible, which is consistent with the critical spatial frequencies.

It is better if the thickness of the film supplements the aforementioned information. In many cases, the thickness of the film can be presumed to be flexible during the replication of the image. This can assist in making a reasonable estimate of the potential error associated with incomplete knowledge of the critical parameters. It is necessary that a model of the crystal lattice be implanted into the software package, which includes the position of the atom and their approximated tenancy. Insertion of appropriate dislocation fields into the prototype of the crystal lattice may aid in building the models with the inherited lattice flaws such as the stacking faults and displacements. The boundaries of the crystal lattice can be replicated by inserting them analogous to the optic axis, and this can be achieved by a transformation matrix in the region associated with the crystal outside the boundary plane.

Best-fit simulations can be achieved through a pictorial similarity of the of a replication of the recorded images. However, it is a more reliable approach to derive a "difference" image wherein the computer program can be employed to simulate the intensity of each pixel and these simulations can then be compared with the concentrations of the through-focus image sequence that has been noted by the CCD camera.

2.6.3 Lattice image interpretation

There is a general supposition that the lattice image presents the position of the atom, which, however, is not correct as well as not an appropriate information.

At the first instance, the lattice image is noted in 2-D and like any magnified image, has a periodic design of variable concentration. The amplitudes that result in variable intensities can be deduced; however, the information associated with the phase is lost. The stage info can be recovered partly from one or several recorded images; the procedure is considered to be relatively uncommon. The literature usually publishes images that have been recorded close to the Scherzer focus. There is no localization of the image interference pattern in the space; it is a sensitive function of defocus pertaining to the objective lens and the thinness of the sample. Additionally, certain other limits are of lesser significance. The conclusion that the model lattice only depicts the atomic positions as well as their occupancy and periodicity only holds as long as the imaging circumstances are well identified and the obtained image form computer simulation is a decent tie.

However, there are certain serious issues that arise mainly because of the nonlocalized behavior of the lattice image. As for instance, it is usually common to superimpose the unit cell of the crystal lattice on the image of the lattice and therefore identify the channels located along the axis of the microscope that have black patches. On the other hand, the observed white patches correspond to the rows of atoms with high atomic numbers along the same axis. Knowledgeable experts will, however, reject the above interpretations. Another excellent example is that related to the apparent structure of the many phase boundaries. The difference in the lattice potentials further complicates this observation. There occurs discontinuity with regard to the refractive index, resulting in a change in the interfering design of one stage with regard to another phase that is usually perpendicular to the plane of the boundary. The deduction of the actual atomic displacements therefore becomes difficult even though there may be the case that such movements can be observed undoubtedly in the obtained image of the lattice crystal.

With the rapid technological advancements, the morphological information can be derived at atomic detail for the lattice image of both the grain and phase boundaries. There is a need to quantify the information and therefore remarkable improvements have been made in the methods to deduce the images through computer simulations, and therefore evaluating the images recorded digitally. In most cases, it has become likely to enhance the model associated with the crystal-like nanostructure that has been detected by lattice imaging by employing a transmission electron microscope.

2.7 Application of TEM for carbon materials

Figures 2.1 and 2.2 depict the selected area diffraction (SAD) patterns. d_{002} and L_c can both be measured using an appropriate internal standard. The 10 and 11 Debye-Scherrer rings are slim, symmetrical as well as intense in the regions where the hk reciprocal lines have been observed to be parallel to the incident ray. The beginning of the graphitization process is marked with the detection of the 112-diffraction line.

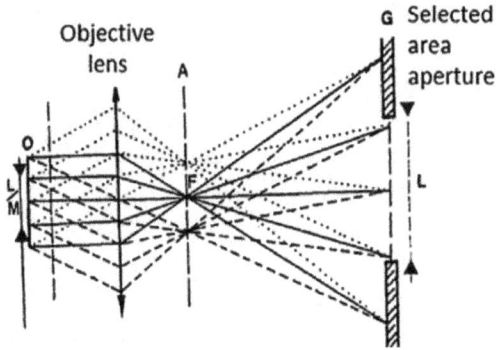

Fig. 2.1: Selected area electron diffraction (SAD) (reproduced with permission from [4]).

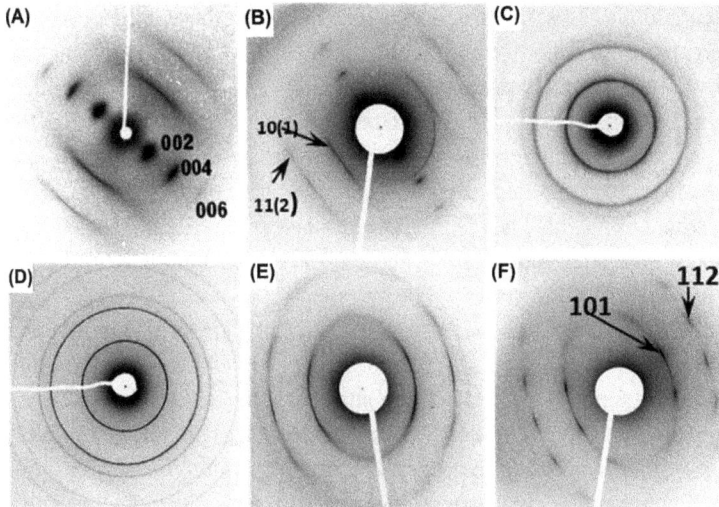

Fig. 2.2: SAD patterns during carbonization and graphitization (reproduced with permission from [4]).

The lattice fringe (LF) mode has been revealed to be reliable in carrying out the study associated with the crystalline domains. A small image analysis can be used to measure the spacing in 002 or 100 LF. Numerous flaws like dislocations, distortions, and piling disorders could well be evidenced. This could also aid in the identification of the atomic structure.

Contrasted bright-field (BF) image, in combination with the successive dark-field (DF) image, results in the localization of the object areas. Their entire quantity is dark in the BF. These are bright in the successive *hkl* DF image. As can be observed from the 002 DF image that when oriented edge-on, the piled-up graphene layers are bright. This is considered to be one of the appropriate techniques to investigate and analyze the disordered or the amorphous states. This is because the percentage of units scattering the lights is lower when compared to other modes. In the case of the 002 DF image, the percentage is as low as 8.5% in comparison to 100% for the case of LF image [4].

A lot of studies can be sourced wherein the two models are opposite to one another. The dominating variability of the as-deposited coating has been often revealed in various studies as one of the primary reasons for the contradictory results obtained while employing the DF technique. The as-deposited films are considered to be sensitive to the incident rays or other external solicitation. As revealed from Fig. 2.3, bright spots seem all over the DF images. This is irrespective of the dimensions and the position of the objective aperture in the Abbe` plane. Since they are insensitive to the through-focus series, these are not localized in the space. The size of the bright spot infers to zero as can be observed from the curve in Fig. 2.3a as the resolution of the image is improved. This could be attributed to the real amorphous state, in which case, the image formed is due to the whole filling of the reciprocal region by the intensity. The state is so unbalanced that an abrupt alteration in the image, as shown in Fig. 2.3b curve B, is observed even if exposed for a certain time to the incident ray. In the SAD pattern, the dots are visible in the folds of the film. It is near $s = 1.9 \text{ nm}^{-1}$ that the halo in the image can be observed. This indicates the presence of coherent scattering domains.

As a result, the focus is now conceivable and the Kakinoki consequences are improved as depicted in Fig. 2.3b. Since there is no initial presence of the coherent scattering domains, the as-deposited carbon coat is primarily in the amorphous state (Fig. 2.3 (c, d)). Observance of a tenacious halo near $s = 1.9 \text{ nm}^{-1}$, due to the localized coherent scattering domains, has been revealed on subjecting the carbon film to a robust current ray and heat treating the same to below 500 °C. The real interlattice plane spacing is one of the reasons for the occurrence of the halo near $s = 1.9 \text{ nm}^{-1}$. Two faint bright dot populations have been observed to be near 4.6 nm^{-1} and 8.7 nm^{-1}. The coherent dispersal areas are observed to be near $s = 2.6 \text{ nm}^{-1}$ when the heat treatment temperature is extended to beyond 900 °C (Fig. 2.3(c)) and 1,000 °C (Fig. 2.3(d)). As a result, the carbon film can be referred to as turbostatic.

Similar information is found in usual samples such as kerogens and coals. However, in the case of immature kerogens, it is in the range of $s = 1.9$–2.2 nm^{-1} that the occurrence of the majority of the bright spots has been evidenced. This forms around

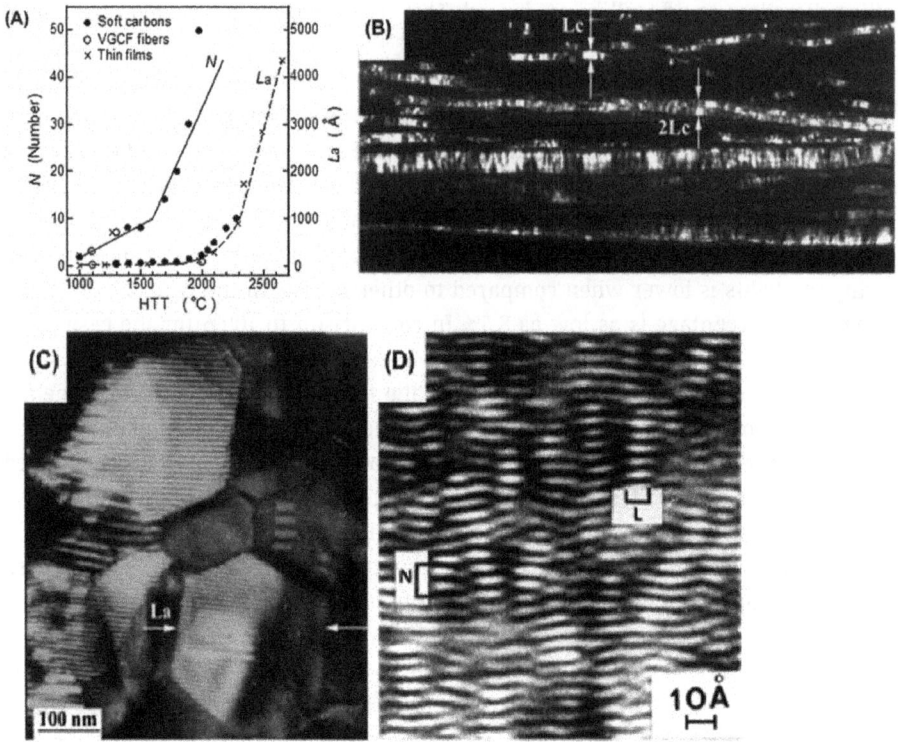

Fig. 2.3: Effect of the objective aperture on the bright dot size and its position (reproduced with permission from [4]).

71–79 mass% of carbon and can be attributed to the evolution of the carbon-enriched kerogens.

The halo in the case of the coal, as there is increase in the carbon content, occurs at $s = 2.2$ nm^{-1} and then a 002 reflection occurs at around $s = 2.8$ nm^{-1}. The later occurrence is due to the sudden alteration in the mass fraction of carbon mass i.e., >85.5%. Below the mass% of 85.5%, the products are rich in heteroatoms [4].

From the description of the data that has been presented through the above discussion, it can be concluded that it is in the case of as-deposited carbon films that there is presence of the amorphous state. However, this is not a stable configuration.

2.8 Conclusion

In transmission electron microscope, the high-energy electrons are dispersed as the electron ray pierce through the sample. Electromagnetic lenses are then used to focus the transmitted electrons and therefore resulting in the formation of the well-resolved image. The resolved mage can then be watched on a florescent screen or noted on a photographic blend or through the charge-coupled device. The presentation of the modern transmission electron microscope has been improved with the incorporation of the field emission cannons.

The wavelength of the high-energy electron ray is well below the interatomic spacing in solids, which provides for atomic resolutions that is not possible to achieve practically. The electromagnetic lens that is used for the image formation in the microscope, occupies a major portion of the total optical path length. As such, the electromagnetic lens suffers from the severe defect that limits the divergence angle of the ray that would have aided in focusing the ray sharply to a small focus. Astigmatism, and chromatic and spherical aberrations are the common lens defects. With the increase in the lens aperture of the objective lens, the diffraction limit on the resolution is improved. On the other hand, deterioration in the aberration limit is observed for the increasing aperture of the objective lens. Therefore, there is an optimal aperture of the objective lens that provides a compromise between the diffraction limit and the aberration limit. The aperture of the objective lens therefore lies between 10^{-2} rad and 10^{-3} rad. However, the value of the objective aperture has been increased with the availability of the spherical aberration corrector. The scattering of the electrons is avoided through the inclusion of vacuum within the microscopic column. Moreover, to safeguard the monochromatic ray, high voltage stability of the electron source is also required. Also, for a suitable operation of the electromagnetic lens, a good current stability is also desired, which is also needed to preserve a steady focus.

The specimens for transmission electron microscope have to necessarily be smaller than 100 nm in thinness. With a thickness smaller than 100 nm, the inelastic scattering of the transmitted ray, on passing through the specimen, can be avoided. Preparation of a good specimen is therefore critical. There is a combinatorial approach that is typically followed for sample preparation and it includes the following: electrochemical, chemical, and mechanical. Ion milling is often followed as a last stage of the specimen preparation wherein the sputtering process erodes the surface atomic layers. This is accomplished by employing an incoming ray, comprising of inert gas. However, it is frequently required to deposit an electrically conductive layer to avoid surface charging of the thin films that have been prepared from insulator materials. Morphological studies can be accomplished for certain specimens through the preparation of slim imitation of the surface of the sample as well as by employing a thin-film transmission sample.

The contrast in the case of transmission electron microscope may be due to any of the following: mass thickness, coherence deflection, and phase contrast. Mass-thickness contrast is the most common in the case of the biological samples and in order to enhance the contrast, a wide diversity of weighty metal stain agent have been established. The diffraction contrast is the most common source of image formation in the case of crystal lattice defects and the reason may be attributed to the fact that the amplitude and the phase of the incident ray are greatly affected by the defects associated with the crystal lattice. Phase contrast successfully images the periodic crystal lattices, particularly when the microscope can be functioned with adequate resolving power. Phase contrast also aids in the detection of variations in the lattice cyclicity at the boundaries, and the same can be interpreted using a suitable software package.

The lattice or phase-contrast image is referred to as an interfering image wherein there is a combination of several coherent-deflected electron rays in the plane of the image. There are phase shifts that are experienced by the electron rays that have trailed different paths while passing via the field of the electromagnetic lens. The phase shifts rest on the electro-optical possessions of the electronic basis and are described using the contrast transfer function. There is no dependence of the specimen thickness as well as the focal plane of the image on the periodicity of the lattice image. However, the aforementioned parameters lead to gross differences in the intensity of the noted image. The phase shifts can, however, be compensated by taking image at an under-focused image below the Scherzer focus. Computer simulations have been used widely to model the lattice assembly to an accuracy that is superior to the portion of interatomic spacing.

Chapter 3
Scanning electron microscope

3.1 Introduction

The scanning electron microscope (SEM) provides images of samples approximately similar to what is expected by the human brain and the eye, physiologically. This is because the resolution of the image produced by the microscope is superior to the spatial resolution of the view field. This is the way that the eyes and the cortex have evolved over the years to recognize the three dimensions of the actual world. The flatness revealed in the case of the optical and transmission electron microscope is replaced by an image in the SEM that seems to be very analogous to the show of light and to the shadow above the valleys in a landscape. Such characteristics are observed as dips and protrusions as part of a three-dimensional object. The optical illusion, as depicted above, can be accomplished through two additional features. The first feature relates to the depiction of the location in the image, detailing the depth, particularly perpendicular to the plane of the image. Stereoscopic imaging is a technique through which an SEM can provide info about the depth by noting the images from two distinct viewing positions. Another feature that is absent in the images taken by a scanning electron microscope is color. Color coding can be employed to aid the human eye in recognizing the color. Color might be employed to improve the contrast, especially while comparing two images, while data is processed. The crystallographic info in a morphologic image can be ensured through a suitable color-coding process [2].

Scanning electron microscope has gained popularity owing to its potential to reveal detailed information of those that are dislocated along the optic axis. Moreover, a two-dimensional view of the image is produced at the level of the image. The use of scanning electron microscopy has extended to different branches of science and engineering since its introduction in the 1950s. A description is first made from the several imaging signals that are easily traceable in a scanning electron microscope and arising from the interface of the focused ray of the electrons with the solid specimen. The chapter reveals evidence on the interpretation of the information from the signals in tandem to the especially available techniques to enhance the contrast of the image and assist in the interpretation of the image [2].

3.2 Components of a scanning electron microscope

The elementary assembly associated with a scanning electron microscope was described in Chapter 2. A comparative analysis was also made with the transmission electron microscope in Chapter 2. Following are the main components of a scanning

https://doi.org/10.1515/9783110997590-003

electron microscope instrument: microscope column, computer software, hardware and signal detector systems, and the recording and the display systems.

The microscopic column is kept under vacuum. Like in the case of the transmission electron microscope, the vacuum system and the air lock for the specimen are integral parts of the scanning electron microscope. The vacuum requirements are more stringent in case the electron source is the field emission gun. A specialized chamber is also required at times when the sample to be analyzed is very large or the specimen requires to be observed below cryogenic conditions. There is a wide availability of different types of specimen stages; however, the steps that are required to prevent contamination of the column owing to outsized samples or special stages need additional features.

The electromagnetic lens of a scanning electron microscope is an inversion of the objective lens employed in the transmission electron microscope. Unlike the magnified image of the specimen produced by the transmission electron microscope, a reduced image is produced onto the plane of the image by the scanning electron microscope. The design of probe lens also differs in both the instruments. The probe lens, in the case of the scanning electron microscope, is required to possess appreciable working distance owing to the fact that three-dimensional substances are viewed in the scanning electron microscope. The working distance is one of the important parameters in the operation of the scanning electron microscope; it might vary from 1 or 2 mm to 50 mm or even more. The microscopic column, in the case of the scanning electron microscope, is required to be equipped with electromagnetic scanning coils because the electron ray probe must scan over the specimen in the x-y plane. The electromagnetic scanning coils are positioned over the probe lens in the scanning electron microscope [2].

Most of the finder sensor setup for the scanning electron microscope is built into the microscopic column. The detector for optical fluorescence is the only exceptional detection accessory. Some of the general finder sensor components include the high-energy electrons, energized x-rays, low-energy electrons, etc. There are two distinct variants of X-ray detectors: wavelength-dispersive and energy-dispersive. In the case of energy-dispersive spectroscopy, the chemical composition of the sample can be determined through the analysis of the energized photons that are composed as a function of their energy. In the case of wavelength-dispersive spectroscopy, the concentration of the energized X-radiation is analyzed, which is a function of the wavelength. Energy-dispersive spectroscopy detectors are highly efficient as they cover a wide range of the sample simultaneously. However, at times, there may be an overlap with the typical peaks in the X-ray finder; they are created by distinct chemical ingredients due to the restricted energy resolution possessed by the energy-dispersive spectroscopy detectors. On the other hand, wavelength-dispersive spectroscopy detectors are required to rotate about the X-ray goniometer so as to scan across the scope of wavelengths that have been composed by the employment of curved crystal detectors. The processing can be time-consuming because of the limited scope of the wavelength sheltered by an only finder crystal [2].

The recent scanning electron microscopes are equipped with computerized data attainment and storing setup and have replaced high-resolution cathode ray tubes. The microscope system comes integrated with data processing and display systems. The images to be recorded in a scanning electron microscope are large and that too at high resolutions and as such, the requirements associated with the computer software and hardware are not trivial. With advancements, the required software for processing of data has been developed and integrated with the scanning electron microscope. The morphological data obtained from the scanning electron microscope and, in particular, the secondary electrons, can be routinely integrated with the compositional information achieved using energy-dispersive spectroscopy. Therefore, the modern scanning electron microscopes have now been expanded with micro chemical capability and hence are capable of identifying the crystallographic positioning of the specific crystalline grains that are present in the microstructure. This can be achieved by the analysis of the patterns obtained from the electron backscatter diffraction of the lattice. The obtained crystallographic info can be mapped onto the image information from secondary electrons and this is obtained through the inbuilt mode of operation in a scanning electron microscope; it is referred to as orientation imaging microscopy. Thus, the potential of the scanning electron microscope to integrate chemical, morphological, and crystallographic information has expanded its application in research and industrial platforms [2].

A two-dimensional stereological analysis of the section of a sample aids in the determination of only a limited number of parameters associated with the microstructure. Deduction of additional three-dimensional information from a 2D section of the sample, as for instance, the grain size distribution, can be misleading. Serial sectioning is one of the probable answers to the aforementioned limitation. Under such a process, the sample is sectioned sequentially at regular intervals and the data that is recorded is united to develop a three-dimensional model for the morphological analysis. One of the important variants that is now available with the scanning electron microscope is the dual ray focus-ion ray that successfully images at high resolution, the serial sections of a specimen. With technological advancements, practical and three-dimensional microscopy is increasingly available.

3.2.1 Interaction between the electron ray and the specimen

The electron ray undergoes both elastic as well as inelastic dispersion on penetrating the solid sample. In the case of a thick sample, it is the inflexible dispersion of the electron ray that predominates and eventually reduces the energy of electrons in the electron ray to the kinetic energy of the specimen i.e., kT. There is a complexity associated with the various processes that happen along the dispersion pathway of the electron. There will be lesser ambiguity associated with the interpretation of the image contrast if there is a proper understanding with regard to the signals generated as a result of the different interactions occurring between the ray and the specimen [2].

3.2.2 Conditions associated with ray focusing

The characteristics possessed by the probe lens used in a scanning electron microscope are similar to that of the objective lens used in transmission electron microscope. The finest achievable resolution cannot be improved more than the probe size being focused on a specimen. The positions of the electron source, probe lens, and condenser setup effectively reverse the path of electrons in a scanning electron microscope when compared to the configuration in a transmission electron microscope. For instance, the electron source in the scanning electron microscope would be in the position where the image is created in a transmission electron microscope. The apparent dimensions of the aperture are reduced by the condenser system in the scanning electron microscope as opposed to the magnification of the image in the transmission electron microscope. The probe lens in a scanning electron microscope forms the ray probe in the image plane wherein the source would be in a dimension of the transmission electron microscope. The reduced image of the electron source is related to the electron ray probe [2].

There are three major restrictions with the least span faced by the probe ray in the specimen plane: the chromatic and sphere-shaped aberration of the probe lens. Due to the greater significance of spherical aberrations, there is a maximum beam current limit for focusing into a probe of a specific diameter. The extreme current is a robust objective of the electron cannon being used. It is also the critical factor for the preference toward the field emission gun. The third limitation relates to the available working space below the probe lens. Samples, typically of 20 mm diameter, are preferred; however, they can range from 2 to 50 or even 100 mm in several implementations. The specimen size that can be worked upon in a transmission electron microscope is restricted to 3 mm [2].

In practical situations, the ray current is one of the serious causes of concern because of the fact that the ray current varies as the cube of the diameter of the ray. A poor signal-to-noise ratio may result owing to a reduction in the ray current and therefore, it is a serious issue. The field emission sources have the potentiality to produce electron probes with a very high density of current. The operating span between the specimen and probe is another critical factor. Classically, a tradeoff is required to accurately adjust the operating span. The required resolution aids in fixing the minimum probe diameter. This also identifies the extreme depth of the arena, finally translating into the extreme operating span for the samples below the probe lens [2].

The minimum size required of the primary electron source has been drastically reduced with the marketable outline of the field emission guns. As such, the available density of the current in the probe is also increased by four orders in magnitude and therefore a probe of a much reduced size is possible, which has led to using an electron ray of lower energies despite keeping the signal-to-noise ratio within acceptable limits. The finest resolution that is now available is 1 nm and is presently accessible for ray energies down to 200 V. The order in size as well as the energies can be compared with the generation of the signal at energies below 5 kV or achieving a secondary image resolution that is better than 20 nm [2].

There is also criticality and importance related to the spatial distribution and temporal stability of the electron ray. The experimental approach in which the spatial distribution might help is by removing the blade brim across the Faraday cup gatherer. The electron beam diameter is defined as the width of the experimental current distribution measured at half the maximum recorded current and is referred to as the full width at half maximum (FWHM) diameter. However, this is reflective of the object controller to the entire current. Distribution of the current usually comprises a lengthy end, extending from the chief point, and resulting in a high noise-to-signal ratio in the background. This is, however, unimportant in the case of many imaging purposes. The X-ray data acquisition is, however, not affected significantly, particularly when there is variation associated with the concentration across a stage edge, and is identified quantitatively [2].

The reduction in the size of the probe is one of the first requirements that can aid in reaching the ultimate resolution in a scanning electron microscope. The image resolution is also significantly influenced by the volume of the material excited within the specimen, which generates the signal being collected. As an example, if the diameter of the focused probe is only 2 nm and if the image is generated from the backscattered electrons, originating in a region that is 500 nm in diameter, then the resolution of the backscattered electrons is determined by this larger dimension. Although the performance of the scanning electron microscope may be determined by the size of the probe, this is not considered as the characteristic associated with the resolution of the collected data pertaining to the image. This is totally different in comparison to the transmission electron microscope, wherein the resolution is identified by either the point resolution or the info limitation. These are features of the electro-optical setup irrespective of the specimen [2].

A brief discussion is now provided on the scanning setup. The information is obtained by scanning the electron probe over the specimen's outer layer and then gathering the resulting signal. The concentration of the probe, the efficacy with which the signals are recorded, and the speed of scanning also influence how rapidly the data is collected. A slower scanning speed is required for a weak signal in order to have improved signal-noise ratio of the image. When gathering X-ray data, the detection limit is ascertained by considering the statistics of the collected data, especially in cases with low inelastic scattering cross-sections, along with the efficiency of data collection. The accuracy of the analysis is therefore greatly affected under such conditions by the stability of the ray current or by the current drift [2].

3.2.2.1 The energy loss and inelastic dispersion

Monte Carlo simulations can be used to quantitatively simulate the calculations associated with inelastic dispersion paths of the electrons. The process ignores the crystallographic dispersion effects such as the effects associated with the lattice anisotropy and the process of channeling, wherein the electrons are scattered into the preferred crystallographic directions. An irregular dispersion path is followed by the electrons in the path of the ray propagation, which loses energy as there is an increase in the length of the crystal. It is impossible to determine the average trajectory of the multiple electrons. However, it is likely to quantify the dual perilous factors i.e., the penetration depth in the specimen and the envelope defining the boundaries of the electron trajectories. This can be accomplished for electrons that have energy exceeding the average value. The diffraction depth can be defined as the depth beyond which the electrons are scattered randomly and therefore, there is all likelihood that the electrons at this depth might be moving in any direction within the specimen. The electrons can continue to diffuse as the depth increases beyond the diffraction depth. The diffusion taking place is such that the dispersion angles are self-governing of the direction. The diffusion depth resembles the specimen thinness if electrons from the incident ray penetrating the thin film sample are collected in a Faraday cup. The critical sample thickness is defined as the thinness that lessens the current of the conveyed ray to one-half of the primary value. The infiltration depth is defined as the depth at which the energy of the electron diminishes to one-half of the thermal energy i.e., kT. Regarding Faraday the cup experimentation, the infiltration deepness can be defined as the thinness of the sample at which the current associated with the transmitted electron ray is zero. The Faraday cup is also expected to gather certain secondary electrons and thereby results in an increase of the transmitted current. However, in general, the experiment works well. With the increase in atomic number, both the depth of penetration as well as the depth of diffusion decreases. The same is also true with the decreasing energy associated with the electron ray. The change in the form of the encompassing electron path with the ray energy is almost similar; however, the same is not true with the change in atomic number. The envelope's shape delineates the trajectories of scattered electrons with a specific average energy, and it undergoes substantial changes depending on the atomic number. The reason for the same can be attributed to the proportional variation of the lateral spread of the ray with the difference between the depth of diffusion and the depth of penetration [2].

3.2.3 Excitation of electrons associated with the X-rays

Ionization occurs in case the energy of the incident electrons exceeds the energy that is required to knock out the electrons from the atoms present in the sample. The ionization process results in an inelastic dispersion that in turn diminishes the energy associated with the incident electron ray. The energy reduced is representative of the ionization energy as well as by the amount the energy of the atom in the specimen that is enhanced. The atom's energy may subsequently decrease as the electron moves from a higher energy state into the vacant position within the atom. The shifts are always conveyed by the release of photons. The emitted photons have energy lying in the region of the X-ray if the energy of the energized state of electron resembles the expulsion of electron from the inner shell of the atom [2].

The decay of the energized atom to its lowest energy state takes in succeeding phases and is accompanied by a release of the photon. These photons have associated energy and wavelength, and each photon is associated with a single step during the shift of the energized atom to its lowest energy state. For the ionized state to be reached, the energy of the incident electron ray in the inflexible component is constantly required to exceed the threshold energy that is required for the formation of the ionized state. On the other hand, the energy corresponding to the most energetic emitted photon should be much lesser than the aforementioned threshold energy required for excitation. As the atomic energy of the atom is increased, the energy required for ionization also increases owing to the reason that the electrons are closer to the nucleus and as such are more deeply embedded within the atomic structure [2].

A range of wavelength is associated with the X-ray spectrum generated because of the electron ray incident onto the specimen. The wavelength generated can be determined using the de Broglie relationship i.e., $\lambda_0 = hc/eV$, where h is the Planck's constant, e is the charge on the electron, c is the speed of light, and the accelerating voltage of electron is defined by V.

The fingerprint of the chemical element is therefore constituted in the wavelength that corresponds to the characteristic X-ray excitation lines emerging from below the probe. This is therefore one of the powerful methods that aids in a suitable identification of the chemical ingredients and their 3-D spreading within the atomic structure. Since a group of spectral lines are emitted from below the probe, there may be chances that these may overlap partially and thus the chemical constituents associated with them.

The X-rays that are produced by the incident electron ray also have a captivation spectrum and are elevated energy photons that have the potential ability to excite the atoms in the specimen to an advanced energy-ionized state. This is particularly true in the case of short wavelength X-ray photons. As for instance, the K photon that possesses higher energy will be able to excite an atom of inferior atomic number to the K state as the K photon will have sufficient energy to do so. As a result, the captivation of the photon with a higher energy takes place. The atoms in the energized state will then decline backward to a lower energy state and finally to the ground state. This process in turn generates a novel and lower energy photon, which is representative of this second atom. The mechanism is referred to as X-ray brightness. The captivation coefficient μ characterizes the X-ray absorption process and is dependent on the wavelength of the X-rays and the atomic number of the chemical ingredients. The perilous captivation edges in the X-ray captivation spectrum corresponds to the photon energies, which have a threshold for the excitation of the bright radiation [2].

Therefore, the information with regard to the chemical constituents of the sample can be well derived using, together, the excitation and captivation fields associated with either the electrons or the X-rays. The generated X-rays below the focused electron probe emanate from a solid sample or volume of sample, and is well defined as the envelope of energies associated with the electrons that have the potential to produce excitation in the representative radiation toward any of the chemical constituents within the specimen. The size of the bulk element diminishes as the voltage associated with the ray is reduced. This in turn results in an increment of the potential resolution of the X-ray signal. At the same time, the intensity of the emitted signal is also reduced. No characteristic X-rays are generated when the incident ray energy, which is identified by the rushing voltage, is reduced and drops below the threshold energy for excitation. Therefore, a compromise is required to be stuck so as to ensure that significant characteristic X-rays are generated and remains limited in the area of specific attention. The concentration associated with the X-rays will increase with the increasing electron ray energy beyond the least energy that is required to excite the distinctive finder. As the energy associated with the electron ray exceeds to over four times the critical excitation energy, there occurs a decrease in the concentration of the X-ray signals that escapes from sample. The reduction in intensity may be attributed to the growing captivation of the energized X-rays in the specimen because of the generation of X-ray signals below the sample's surface. This is produced as a consequence of the increasing depth of dispersal with the increasing energy of the electron ray. Therefore, there exists an optimum energy for the electron ray that can provide extreme excitation to the X-rays. It is a usual practice to select the energy of the electron ray that corresponds to the wavelength of the smallest representative X-ray that is required to be spotted [2].

The efficacy associated with the X-ray creation is small and there is emission of X-rays at different angles. A greater amount of the X-rays get trapped in the sample or do not reach the intended detecting system. This is the reason that there is inefficiency in the collection of X-rays. A solid-state finder is employed to promote the energy-dispersive collection of the X-rays. Another alternative is to provide a crystal spectrometer that uses a sequence of differently curled crystals and are characteristic of the wavelength of attention. Excellent spectral resolution is one of the advantages associated with the curved crystal diffractometer. The excellent resolution aids in the reduction of the chances of ambiguities associated with the peak overlap of the X-ray wavelength [2].

The capability to discriminate energy by the cryogenically cooled semiconductor crystal lays the foundation for the solid-state X-ray detectors. The generated electrical charge by the photon in the detector system varies proportionately with the energy associated with the incident photon. The generated charge results in the current pulse, which in turn varies proportionately with the energy absorbed by the photon.

One of the frequent issues that is plagued by the energy-dispersive spectrometer is that associated with the overlay of the representative crests of the signals. The afore-mentioned issue can be solved by the selection of a single or extra characteristics re-lease line for study from the spectrum of the available X-ray energy. A microscope provides a shield to the detector and prevents it from being contaminated. This is achieved by employing a polymeric film window and the employed polymeric film is transparent to all wavelengths except the longest wavelength i.e., the lowest energy X-rays. The detectors have the potential to detect photons whose wavelengths do not exceed 5 nm, which corresponds to the emission of X-ray from the lightest atomic ele-ments such as carbon, oxygen, boron, and nitrogen.

All the energy-dispersive spectrometer detectors are limited in their efficacy – the rate at which they accept the photons. This in turn resembles the period essential for decay of each charge pulsation in the finder system. The decay time, also referred to as dead time, is classically less than 1 microsecond and through this period, the photon counts are not recorded reliably. The count rates are not expected to exceed 10^6 s^{-1}. Inevitably, some of the photon counts may be lost owing to the fact that the photons are generated randomly. The counting system registers the proportion of the dead time between the photon counts. The order of the acceptable dead time is typi-cally 20%. A lower value of the dead time signifies that the rate of data accumulation is lower. There are three distinct formats for depicting the X-ray signal: X-ray spec-trum, X-ray line-scan, and the X-ray chemical concentration map.

The X-ray spectrum is employed primarily for the identification of the chemical elements that are characteristic of the X-ray fingerprints. The X-ray spectrum might compose with the condition when the ray is immobile and at a definite position on the surface of the sample. The spectrum might collect when the ray scans a particular zone and this technique may aid in preventing the buildup of contamination within the microscope system. The characteristic gesture gathering time needed to assure proper recognition by the energy-dispersive spectrometer of all the constituent ele-ments in the specimen is 100 s.

In the case of the X-ray line-scan, the ray is passed through a particular area of the specimen. This is accomplished in separate stages and a prerecording of the signal is done at the individual stage. The number of stages per unit span of the line deter-mines the 3-D resolve for X-ray analysis. The volume element within the sample limits the best achievable spatial resolution; while on the other hand, the time for gesture attainment at individual stage defines the limit of the chemical constituent limit. The generated signal can take the form of a complete energy-dispersive spectrometer spectrum. Alternatively, there may be a feasible option to select a number of charac-teristic X-ray energies. A significantly higher digital storage may be required for a full-spectrum mapping, in comparison to that required for a single line scan. In either of the mechanisms of data acquisition, the intensity of the recorded X-ray crests are presented as a function of the location of the electron ray. The choice window that corresponds to the spectrum of wavelength of interest aids in excluding certain set of energies and thus the associated photons. For each position along the beam traverse, the number of counts corresponding to the chosen characteristic energy peak is displayed. The aforementioned method of function aids in the determination of the concentration gradients and in the analysis of the segregation effects associated with the grain boundaries, interfaces, and the phase boundaries [2].

The incoming ray of electron is rastered over the sample's selected zone of interest in an X-ray chemical concentration map. Subsequently, photons are gathered within one or more energy windows that are distinctive for the chemical constituents of interest. The color-coded dots display the photons detected for the corresponding X-ray stroke for position on the screen as the ray is glance over the sample surface for analysis. The range of wavelengths that is of particular interest is chosen in the energy gap that resembles the representative X-ray photon energy. There is the added advantage to view the several detected elements simultaneously, and each such map corresponds to the distinct energy window that uniquely code for diverse constituting components. In many of the scanning electron microscopes, the area that X-ray plots might gather autonomously. The count periods might get accustomed as per the requisite degree of importance. It is also possible to rectify the background for each of the recorded points and excitations, corresponding to the secondary fluorescent. It is then possible to quantitatively analyze the chemical constituents of the specimen surface. However, the time consumed is very high during such a process wherein enough ray and sampling step equilibrium is collectively ensured through the conditions, ensuring least contamination of the specimen surface. As such, to successfully implement the aforementioned process, high instrumental standards are required in addition to the experienced operational competence [2].

In most of the scanning electron microscopes, the above three means of operations include the prolonged electron probe radiation of the specimen surface. Thus, a higher dose of irradiation accompanies the acquisition of the necessary statistics associated with the photon counting process. As a result, the specimen surface gets deposited with the carbonaceous contamination film. It is always preferential to keep the detection time minimum so as to statistically collect a significant proposition of the signal. The X-ray spectra can be affected significantly by the deposited carbonaceous contamination film by adding to the irradiation and also by changing the conditions associated with the acquisition of the data and hence its analysis. Opting to scan the probe across the area of interest is thus a favored choice compared to a point-wise analysis. This approach enables the collection of a similar number of photon counts from a specific coordinate location, accounting for both electronic drift and mechanical specimen step motion. One of the primary reasons for achieving statistical significance associated with the data acquisition process is the overriding considerations in the recording of the X-ray data. One of the simplest assumptions that can be made is the obeying of the Poisson distribution by the tally data by the individual points of the specimen surface. The counting error related to representative X-ray line for every component is $1/\sqrt{N}$. Here, N is total number of tallies noted for the chemical component from the volume element of specimen under analysis and being energized by the high-energy electron ray [2].

Based on the preceding discussion, it is evident that both point analysis and the comprehensive spectrum collection yield data related to either the fixed beam coordinate or a chosen scanned area. The statistical significance is high in the case of moderate counting and can be determined by first making corrections in the associated contextual tallies. The contextual modifications are quantified in two distinct gaps located on each side of the characteristic X-ray line of interest.

Reduced statistical significance is associated with the results obtained using line analysis and varies as the square root of numeral of the image basics or pixels located lengthways along the stroke chosen for examination on the specimen surface. This method is preferred because it aids in the determination of the concentration changes near the interfaces, surfaces, and the grain boundaries. Such concentration changes may be, for instance, the exclusion effects and the dispersal absorption slopes related to precipitate kinetics of the second stage of the specimen [2].

Far lengthier count times are required in the case of elemental maps and are considered to be in a different class altogether. Only the appropriate backscattered electron picture may be used to get the elemental maps. A decent secondary electron picture with extreme resolution will have in excess of 10^6 resolved pixels. On the other hand, a prohibitive counting time is required in the case of an X-ray rudimentary plot at a similar resolution. As such, it is not at all possible and is therefore impractical. This is because neither the machinist nor the firmness associated with the electron probe are closer to aforementioned prohibitive time [2].

The pertinent question that arises at this juncture is whether elemental mapping can be considered to be a good idea. What are the methods that can be practically adopted to deal with elemental mapping? Let us assume that the counting time is 100 s and the requirement associated with the statistical significance is 95% confidence level. Then, the tally amount that is required for a resolved X-ray pixel would be 1000 in the mapping of the X-ray image. If even a comparatively high tally rate of 10^4 s^{-1} is considered, then the calculation suggests that the corresponding acquisition rate for the formation of elemental map will be 10 pixels per second. Hence, it would require approximately 25 min to acquire an elemental map of the sample's surface with 128 × 128 pixels and hence more than a day to acquire an elemental map for a significant surface area with the pixel density approaching as close to that of the secondary electron image. The 3-D resolution of an inferior electron resolution continues to be preferred over that produced using the elemental X-ray map, which may simply be because of the differences associated with the scope of the volume components in the specimen from which the X-ray emissions are emitted as well as the secondary electrons that take place. In fact, there is no sense in generating the elemental map that has the same bulk of quantified points as that of the image being produced using secondary electrons [2].

The spatial resolution associated with the line scans as well as the elemental map is limited by the dimensions associated with the X-ray producing area under the probe, and this is independent of the problem of statistical counting associated with X-rays. The electron dispersal span in the specimen determines the size of this zone. The dispersal span depends on the atomic number and the power of the ray or on the density of sample. Typically, the diffusion distance ranges from 0.5 μm to 2 μm. If in case the value of diffusion distance is chosen to be 1 μm, considering the maximum number of X-ray counts, it would take 16 min to obtain an elemental map from a 100×100 μm² region, which is approximately 10^4 pixels. The elemental maps can be obtained from a larger region in case the counting times are increased significantly; however, this seems to be impractical. The impracticality may be associated with the fact that imaging smaller areas with advanced amplifications will be outcome in capturing less image points also such as blurred images, which may be produced because the size of the excitation envelope dictates the resolve [2].

With the aforementioned discussion that resolve in the basic plot is intrinsically inferior to that obtained using the subsequent electron, the best practice suggests making use of X-ray data. It is possible to observe a detailed morphological image of the specimen by superimposition of the X-ray elemental map onto the image produced using the secondary or the backscattered electrons. As such, the localized variation of the chemical composition can also be made. It is then possible to gather the X-ray data for 128×128-pixel array in less than 30 min. The area determined by the excited envelope of the X-rays represents the area associated with each pixel, wherein the excitation envelope is in the order of 1 μm in diameter. The low-resolution fundamental plots are a very valuable director in the determination of the localized chemical composition variations for the features close to that noted at high resolution subsequent electron or backscattered picture, irrespective of the differences in the orders of magnitude of the density associated with the resolved pixels [2].

3.2.3.1 Backscattered electrons

There will be a portion of the incident high-energy electrons that will be dispersed by an angle that is greater than π and such electrons possess limited chance to getaway from the surface of the specimen. The factors that determine the fraction of these electrons are the mass bulk or more precisely, the usual atomic number of the samples. The backscattered electrons therefore aid in the resolution of the local variations associated with the mass concentration as well as the outcomes into the atomic number contrasts. The portion of the backscattered electrons is, however, less dependent on the associated energy of the incident ray energy and decreases as the energy increases. The atomic number disparity can be quite beneficial since it offers to differentiate well between the different phases of the microstructure with far better resolution in comparison to that achieved using the X-ray microanalysis. At lower voltages, the atomic number disparity is, however, additionally noticeable. The backscattered electrons originate from a surface layer that has a thickness equivalent to the diffusion distance. Backscattered electrons emanate from the region located at a distance within the electron diffusion range beneath the beam and are considerably shorter in range compared to inelastically scattered electrons wherein the following relation holds true: $E > kT$ [2].

The average energy associated with the backscattered electrons is always less than the energy associated with that of the primary electron ray. The scattered back electron, on the other hand, may be used to retrieve crystallography data. The scattered back electron may be observed across a large range of angles in the annulus area, which is near the probe lens's pole sections. The intensity associated with the scattered finder reductions is proportionate to $cos^2\alpha$, where α is the angle between the path of the scattered back electrons and the incident ray. Diffraction of the scattered back electrons is possible by the crystalline specimen, which has a negotiating effect on the signal intensity when the value of α is low, which satisfies the Bragg condition. The diffracted process adds to the intensity at higher values of α, which gives rise to the backscatter that is similar to the Kikuchi line deflection detected in the case of the transmission electron microscope. A charge coupled device can be used to collect the diffraction pattern created by the EBSD patterns. An appropriate computer software package can then be used to analyze the obtained patterns and provide the information associated with the positioning of the crystal at every pixel. Tilting the sample makes the diffusion distance associated with the high-energy incidence electron ray close to the free surface and hence improves the efficacy of collection for the backscattered electrons. An appropriate software package can then be employed to rectify the captured image for potential foreshortening caused by the angular variation between the microscope's axis and the surface normal of the specimen. The information obtained from the EBSD can be analyzed through the color coding for a particular range of crystal positioning. The current associated with the backscattered electrons is only a fraction of the current associated with the incident ray and this covers the advantage associated with the backscattered electrons. The efficacy of the data collection statistics can be improved greatly with the use of the field emission sources [2].

3.2.3.2 Image contrast associated with the images from the backscattered electrons

There are two major sources of contrast generation in the backscattered electron image. Enhanced signals are produced by the sample surface that is slanted to the finder system for the backscattered electrons. However, as the surface is slanted farther from the finder system, the intensity of the signal is reduced. The topographic image of the surface can then be obtained using the segmented annular detector. The signals collected from the segmented detection system that is located on the opposite side of the sample surface are deducted and then enlarged. The topographic contrast associated with the region is enhanced for the areas of the sample that are slanted in the opposite directions and at the same time, the contrast is neutralized due to the differences in the density and more particularly, the atomic number. The collection of the scattered back image from the finder system that covers the probe lens pole section results in minimization of the contrasts associated with the surface topography. More equivalently, the summation of all the signals from the segmented detection system also leads to the aforementioned result, owing to the fact that signals are composed from all probable azimuth angles. The contrasts that are observed in the image are mostly associated with the atomic number, which truly mirrors the variation in the bulk of the specimen. The density variations in turn reflect upon the compositional variations in the specimen as well as in the regions of fine porosity [2].

The resolution of the obtained image by the backscattered electrons is usually higher than that obtained using X-ray in the elemental map. However, the resolution is not better than that obtained using the secondary electrons. The resolution is limited to a range from 50 nm to 100 nm, owing to the direct relationship between the high-energy diffusion distance and the image obtained using the backscattered electrons. The useful information provided by the backscattered electrons corresponds to the distribution of the visible phases on the highly polished specimen surface. The present available commercial scanning electron microscopes combine the information from the backscattered electrons with that obtained using the EBSD and X-ray fundamental plotting. This is one of the outstanding accomplishments that is acknowledged wholeheartedly by the community of material scientists.

3.3 Scanning electron emission

The emission of subsequent electrons from the specimen surface results in a major proportion of the electron current that is emitted from the specimen, which is energized by the top energy electron ray. The coefficient of the subsequent electron release defines the number of secondary electrons that are released from the incident ray of the high-energy electron. The value of the secondary electron emission coefficient is always greater than 1, and may reach several hundreds. The subsequent electrons can be separated into two components: the subsequent electrons that are created as an outcome of the high-energy electron ray entering the specimen surface and secondly, the subsequent electrons that are created as an outcome of the scattered back electrons that come back to the surface of the sample after numerous dispersions, characteristic of inelasticity. The previous gesture emanates from the surface that has a surface area of an order equivalent to the crosssection of the ray probe and the information that is contained is resolved only by the diameter of the probe. The secondary electrons come from the same area that emanate the backscattered electrons. The image can be resolved in a feature on a gauge that is analogous to the resolve associated with the backscattered electron signal [2].

There are four features that hinder the release of the current associated with the secondary electrons from the specimen surface: (1) work role of the sample surface, which can be defined as the energy blockade that is required to exceed the electron located at the Fermi level within the hard, and escape from the specimen under analysis. The work roles typically are of a limited eV. The work role depends on the atomic packing as well as its composition. The work role is sensitive to the contamination and the adsorption film. The deposition of the carbonaceous film on the surface of the specimen is sufficient to obscure the positive effect associated with the work role of the sample. (2) The energy as well as the current associated with the incident ray. More subsequent electrons are formed under the probe area as there is an increase in the energy of the incident ray. However, the number of subsequent electrons that get away from the specimen is lessened because of the generation of an inelastic dispersion of the electrons at a depth further below the specimen surface. With the increase in the energy associated with the incident ray, it is expected that there will be an increase in the generation of the secondary electrons. The increase in the secondary electrons is continuous and increases through a shallow maximum and then decreases slowly as the energy of the incident ray ranges from 5–10 kV. The current associated with the secondary electrons is directly proportional to the current associated with the incident electron energy, and decreases with a decrease in the accelerating voltage of the incident ray. (3) The yield of secondary electrons is significantly affected by the density of the sample. The backscatter electron coefficient R is high for advanced standards of the atomic number Z and therefore, with increasing value of atomic number additional subsequent electrons are generated by the scattered back electrons. Simultaneously, a smaller diffusion distance is associated with the secondary electrons in materials with higher atomic numbers. On the other hand, at a given intensity of the electron ray, with the increasing value of the atomic number, the inelastic dispersion of electrons is more for materials with higher value of atomic number, in comparison to materials with lower values of atomic number. Therefore, it may be deduced that for a specified excitation energy, a large number of subsequent electrons are gathered from a specimen with higher atomic numbers. The dependency of the subsequent electron yields on the atomic number becomes further noticeable at lower energies of the incident electron ray in which case, the diffusion distance associated with the incident electron probe turns analogous with the mean free path of the subsequent electrons. (4) The surface topography is one of the dominant factors in the generation of the contrast effects in the image generated by the secondary electrons. More particularly, it is the localized curving of the specimen surface and the angle associated with the electron probe that greatly affects the contrast in the generated image. The probability of the generated secondary electrons to escape from the solid increases with changes in the localized curving of the surface. On the other hand, the path length associated with the incident ray is determined by the incident angle of the probe. The chances of escape for the secondary electrons from the region increase with the protrusions from the surface of the specimen, which is the surface with positive curved radii. On the other hand, the recessed region, that is the region with negative curved radii, has probabilities of the subsequent electrons getting trapped and hence their chance to escape from the surface is reduced. The application of the bias voltage at the collector aids in the collection of the secondary electrons and therefore it can be concluded that even though some of the regions are out of sight of the incident electron line, the secondary electrons will be able to provide an image of the uneven surfaces, considering both the outstanding contrast as well as the high resolution. Dramatic improvements have been made with the introduction of field emission guns. The resolution obtained is further enhanced by reducing the working span down to a few millimeters and by placing the subsequent electron indicator within the magnetic field of the pole pieces. Such a field traps the subsequent electrons that have been released in the onward track, owing to opening of the inflexible disperse events. On the other hand, the secondary electrons that are produced by the scattered back electrons at wide angles do not influence the in-lens sensor system that is located within the electro-optical column. The gesture associated with the secondary electrons detected by the in-lens detection system is weaker than the signal associated with the secondary electrons that are normally detected. Nevertheless, this is inconsequential given the beam intensities produced by the field emission gun. The resolution of the secondary electrons associated with the in-lens detection system enhances the resolution of the image to a small extent [2].

There are two major factors that affect the generation of the signal and hence generate some complications in interpreting the image contrast associated with the secondary electron image. These factors are the production of the secondary electrons from the primary electrons and their subsequent escape from the surface of the specimen. The escape of the secondary electrons is restricted by the work function to the extent the specimen under consideration is a plane. This is one of the reasons that the concentration of the gesture reduces to nil for subsequent energies under little eV. The number of events associated with the secondary electrons determines the generation of the secondary electrons taking place in the surface region. As such, the reduction in the voltage associated with the ray enhances the contrast associated with the secondary electron image. Moreover, there is no compromise in the resolution of the image.

The high-resolution scanning electron microscopes are capable of 1–2 nm resolution and this is achieved well by the combination of the in-lens detector with the energies from the electron probe, which is typically below 200 eV. Excellent contrast and resolutions are generated with such combinatorial conditions and this is truly reflected by the varying atomic number and the surface topography. With the low ray voltages, there does not seem to be any problem associated with the conductivity of the surface and also there is no need to provide conductive coating to the nonconductive sample. The penetration of the ray under such circumstances is very limited and the huge yielding of the subsequent electrons neutralizes the electrostatic charge. For such samples, the associated features might be related to the element of the sample and not by the conducting film, which settles over its surface, in order to avoid the phenomenon of electrostatic charging.

There is a problem arising out in interpretation of the image owing to the topographic contrast. The secondary electrons may be trapped in the protruding regions and recessed regions of the specimen, irrespective of the efficacy associated with the secondary electrons. More commonly, the emissions associated with the constructive curve regions may extend some distances, beyond the regions of the top constructive curve. An improved release takes place for a certain distance, corresponding to the electron scope in the specimen i.e., x_R. This may be attributed to the subsequent electrons created by the inelastically scattered electrons, which have the potential to escape from the surface of the specimen. The outcome is similar to the darker zone in the image that is related to the adverse curve, which extends for a similar span outside the darkened characteristic. The observed image may look like that a shadow of the hollows and protrusions that are highlighted, which looks similar to what formed is by the dusky sky in a hill station situated in a country. However, the underlying principle is altogether different because all the subsequent electrons are gathered after different guidelines. This may be deceptive at times once the machine is operated at superior amplifications, which are as close as possible to the resolution limit [2].

3.4 Different imaging modes

Apart from the aforementioned three imaging gestures, there are certain other useful modes of operations associated with the functioning of the scanning electron microscope. Here, in the current discussion, two alternate modes of operation have been discussed: electron ray image current and cathodoluminescence.

It is likely to shape an image of the sample that makes use of the current flowing through the specimen, given that the specimen is isolated electrically from the surroundings. This can be accomplished synchronously as the probe is scanned over the specimen. This mode of operation is defined under the electron ray induced current in the following situation. There are numerous variations of such procedures that are successfully employed to examine and analyze the flaws associated with the solid-state semiconducting material. In particular, the defect structure of the semiconducting material is the variable that contributes to its electrical conductivity. The associated electrical conduction may be measured as the electron ray penetrates the surface, irrespective of whether a bias voltage has been applied or not. This elucidates the operational mode of the scanning electron microscope, which is referred to as the electron beam-induced current mode. Under this mode of operation, the entire solid-state specimen might be placed in a compartment and the machine can be operated in situ. The secondary electron signal as well as the specimen current flowing across the inserted solid-state device is modified as and when the electric field is established in the several areas of the solid-state instrument under investigation. It is also possible to image the devices at high resolutions and the mode of operation has been successfully used by the industry to develop devices as well as to analyze the defects associated with the operation of the device [2].

There are numerous optical resources that release electromagnetic radiations when they are suitably energized. The optically active regions of the specimen undergoing scanning electron microscope analysis glow and emit visible light possessing a wavelength i.e., characteristics of the energy heights available at the surface of specimen as a result of being energized by the high-energy electron ray. An optical microscope may be used to observe the fluorescing sample, wherein the firmness of the obtained picture is inadequate for the characteristics of the objective lens of the optical microscope. In another instance, the image of the optically active materials may be observed by the light emitted through the photoelectron detector, which is then amplified, recorded, and then displayed. In this alternative scenario, the resolution is not constrained by the wavelength of the emitted radiation; instead, it is governed by the diffusion distance, which limits the image resolution. Suitable fluorescing stains can be used to label the biological samples and the cathodoluminescence mode of operation may be employed to image the soft tissue structures wherein the different stains may be used to distinctly identify the different features, as each such feature will fluoresce with characteristics of the wavelength within the optic scope [2].

3.5 Preparation of a specimen and the associated topology

One of the most important considerations for the specimen to be investigated using a scanning electron microscope is of the avoiding of electrically static charging of surface of the sample. This is because of the fact that owing to the instability of the associated charge, unstable secondary emissions will result. This in turn will result in a destruction of the image resolution as well as the stability of the image. Charging of a nonconductive specimen surface can be avoided by either working at a less electrical energy or by covering the sample with a thin film of an electrically conducting coating [2].

The specimen must also be accommodated within the specimen chamber, taking into account the limitations imposed by the specimen's geometric characteristics. This is required so as to ensure that the specimen can be tilted at the needed angle with respect to the optical axis and also examine any selected area of interest. As already stated, the chamber in the scanning electron scanning microscope can admit large samples that are above 10 cm in diameter [2].

The specimen is also required to maintain its stability in the void set up and below the high-intensity electron ray. The surfaces are required to be free from any organic residue such as oil or grease, since there presence may result in the formation of contamination-bearing carbonaceous characteristics. This may get deposited on the surface of the sample or on the electrically optical setup or anywhere in the sample chamber. Ultrasonic cleaning must be employed in order to eliminate any loose particle from the surface of the specimen before they are inserted into the microscope. This should be followed by rinsing and then dried in a light heat airstream. The aforementioned safety measures are vital in the case of low electrical energy and good firmness scanning electron microscopes. Secondary electrons are emanated from regions that are very close to the surface of the sample in the case of probes with energies below 1 ke V [2].

3.5.1 Sputter coating

A sputtering unit may be used coat the samples and thereby improve the electrical conductivity of the specimen to result in images with enhanced contrast. There are two types of coatings that can be employed: an amorphous carbon layer or a heavy metal. Heavy metal layers of gold-palladium alloy have been used widely with particle sizes of around 5 nm. Such heavy metal coatings have been observed to interfere the resolutions only at higher magnifications and the contrast in the obtained image has been found to be enhanced significantly. However, the heavy metal coatings are revealed to interfere with the chemical microstudy and is therefore not preferred in case the best firmness is to be achieved. Carbon coatings, on the other hand, can be used for particles with minor unit dimensions, approximately 2 nm. The particle size is under the firmness boundary of the device. Although the carbon coverings do not enhance the image contrast appreciably, the coatings are required for nonconducting samples that are to be analyzed for quickening powers that exceed 5 keV. Reduction in ray voltage is one of the best solutions to take care of the electrostatic charging. However, this is not the solution in case microanalysis is required to be carried out and in cases where the excitation energy associated with the characteristic line of interest is more than a few keV [2].

3.5.2 Failure analysis and fractography

Fractography as well as breakdown study are the main focus of applications using scanning electron microscope. The analysis is not only restricted to metallic alloys but also to application such as engineering ceramics, polymeric composites, metal matrix composites, and semiconductor devices. There are different classes of materials that have been investigated for their morphology using the scanning electron microscope, as for instance textiles, fibers, and natural and artificial foams. The setup, which is unstable at ordinary temperatures and requires cryogenic cooling, has been studied successfully using the scanning electron microscope. Certain important guidelines may be followed in order to extract the maximum possible information from the image generated using the scanning electron fractography. The sample undertaken for the examination must tolerate an acknowledged geometric orientation and spatial, similar to the system or the engineering constituent from which it has been collected. It will be tough to judiciously estimate the importance of any observation associated with fractography in the absence of the aforementioned critical aspect. During the specimen preparation process, utmost attention is required to ensure that the specimen remains intact and there is no alteration to the original system or the engineering component. The specimen must be mounted in accordance with the step x-y coordinates and in accordance with the axes associated with the sample slant. To illustrate, the specimen must be parallel and perpendicular to the path of the crack spread. A lot of time and effort can be managed through such judicious thinking. The process of the probe can be simplified through such an approach. In order to observe significant microstructural features, it is required to record the images over a full range of magnifications. This is because the analysis can be successful only when a proper relationship is established between the surface topography associated with the microstructural feature and that of the observation. If the initial observations are appropriately visible, then an opening amplification in the scanning microscope of 20 times is sufficient. One of the good practices involves scanning over the specimen surface rapidly, experiment with different magnifications, recording a series of images, identifying the characteristic in the amplified images, and relating its geometry with the features that are observable by the naked eye. A factor of ×3 is required to be maintained between successive magnifications as this ensures that nearly 10% of the surface zone is observed at lesser magnifications, which is seen in the subsequent greater-resolution image [2].

3.6 The process of stereoscopic imaging

The scanning electron microscope has the ability to detect and clarify features over a wide depth of arena in a direction identical to the incoming ray's optic axis. Such data may be retrieved by capturing an identical pair of photos at the same amplification and sample space. However, varied degrees of tilting of the sample, relative to the microscope's optic axis, must be maintained. The pictures are referred to as a stereo pair, and they correspond to two views of the sample surface obtained from two separate positional views. If the angle of tilt is correctly chosen, the design is comparable to a stereoscopic 3D display [2].

For a suitable viewing of such stereoscopic images, sophisticated marketable setup is required wherein the picture can be viewed directly through the microscopic system. These commercial systems have never been more popular, primarily because capturing a pair of images can be achieved without disrupting the settings of the scanning coils; instead, it involves tilting the specimen itself [2].

The optimum angle of sample slant that provides an imprint similar to the depth view observed by the human eye is ±12°. This is obtained by the assumption that the microscopic display is observed at some 30 cm span and there is a distance of 5 or 6 cm between the eyes. This is considered to be equal to the amplification depth corresponding to the crosswise amplification of the picture. The sensitivity of the depth will be reduced in case the tilt angle is less than the optimum specimen tilt i.e., ±12°. The impression of the depth is amplified in case the tilt angles are greater than ±12°, which may aid in the recording of the shallow features [2].

The pair of stereo images that have their axis of slant arranged in a straight line and located with the equivalent points of the picture at distances that are equivalent to the separation between the observers' eye, can be viewed with a little practice by most operators, at a normal reading distance. This can be accomplished by simply focusing the eyes on infinity, without the need for additional optical instruments. Commercial stereo viewing systems are often available but not apparent in a two-dimensional recorded image [2].

3.7 Measurement of parallax

Additionally, to influence the vision created by the stereo image, extraction of quantitative information associated with the vertical distribution of the features is also possible. The measurement of horizontal displacement can aid in achieving the aforementioned objective. The distance is measured vertical to the axis of the stereo tilt and for the same characteristic that is observed vertical to the slant axis in the two images encompassed in a stereo tilt. The displacement is referred to as parallax and can be obtained by the following formula:

$$x_L - x_R = 2h \cos \theta \qquad (3.1)$$

where x_1 and x_R are the likely span in the two images that are computed from any static place vertical to the axis of tilt y. The change in height between the two characteristics computed along the optical axis of the incident high-energy electron ray is denoted by h. In calculations, the stereo pair has been considered to be noted at slant angle of $\pm\theta°$ with regard to the normal of the incident ray energy.

The thickness of the surface films can be identified using the parallax measurements. The parallax measurements can also be employed to determine the height of the growth steps as also the displacements associated with the slip displacements. The parallax measurements are also useful in determining the unevenness of the ground and machined surfaces apart from determining the fractal extents associated with the fracture surface.

3.8 Illustrative examples

The importance of voltage contrast in the production of the image was demonstrated for the first time by Chung et al. [5]. This was done for studying the dispersion of carbon black fillers embedded within a polymeric matrix. The use of the SEM technique was to image the SWCNTs embedded in the polystyrene matrix. The general conclusion that could be drawn from the conducted probe is that the SEM technique can be utilized to image the CNT bundles effectively if the imaging technique is employed judiciously. Figure 3.1 shows an example of the imaged multi-walled CNTs, embedded in the epoxy matrix, using different acceleration voltages.

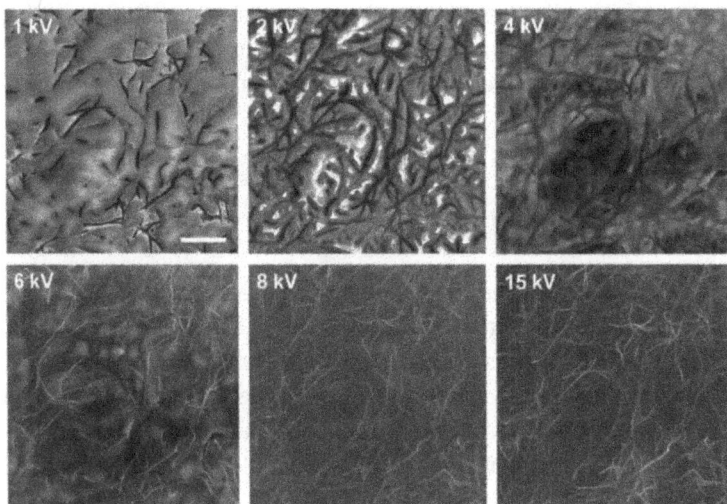

Fig. 3.1: SEM images 0.5% SWCNT(LA)-PI film under different acceleration voltages (reproduced with permission from [6]).

These images clearly demonstrate the usefulness of different acceleration voltages to image nanotubes embedded in a polymeric matrix. The nanotubes generated bright contrast while the polymeric matrix created a dark contrast when the acceleration voltages ranged from 0.2–0.7 kV. The acceleration voltage of 0.7 kV, however, led to the disappearance of the contrast of the nanotubes. The usage of the aforementioned acceleration voltages resulted in contrast inversion at 1 kV, which means that dark nanotubes with bright polymeric matrix were observed. The maximum that the polymeric contrast reaches is around 1.5 kV.

The dependence of the image contrast on the charging mechanism is almost consistent. The following charging mechanisms could be adopted for the various ranges of acceleration voltages: negative charging for acceleration voltages less than 0.7 ke V, positive charging for acceleration voltages between 0.7–2 ke V, and negative charging for acceleration voltages higher than 2 ke V. However, care must be taken to exercise the aforementioned voltage contrast. This is because there are several associated problems that can result in the formation of the artifacts. One such problem may be due to charging history. As for example, if the images were taken in succession, it can be observed that there is an apparent increase in the area occupied by the nanotubes. Moreover, an apparent increase in the diameter of the nanotubes can also be observed. The increased diameter of the nanotubes can be attributed to the distribution of the electrons in the polymeric matrix. Another important challenge is the dependence on the sample type, the type of the instrument, and the type of detectors used to obtain the high-quality images. It is therefore suggested to use high acceleration voltages for specimens with high thermal conductivity and low acceleration voltages for specimens having low thermal conductivity. Moreover, the specimen must be as flat as possible so that the topographic contrast is diminished.

The energy associated with the secondary electrons is low and the maximum escape depth is usually in the range of 50 nm. However, it has been reported in several studies that the electrons may escape from a higher depth as is the case with the CNT/ polymeric composites that have been imaged using the voltage contrast mode. Few research reports have revealed that the secondary electrons can escape from a depth of more than 1000 nm. Studies have also been directed to develop procedures to study the dispersion of nanotubes using quantitative approaches. This has been approached by combining the imaging analysis techniques such as fractal dimension, radial power spectral density, and Minkowski functionals with the images taken by employing the voltage contrast techniques. The quantitative information can be analyzed through the framework that aids in the determination of the persistence length, dominant bundle size, fractal dimensions, and the spacing between the bundles.

Voltage contrast technique can be employed to obtain the quasi-3D image but a true reconstruction of the 3D model from these images has not been reported in the various studies. Figure 3.2 depicts the 3D SEM image of 0.5% SWCNT-epoxy sample tilted at −5° (left) and at −5° (right) angles.

3.9 Conclusion

In the scanning electron microscope, an incident ray with enormous energy electrons is made into a fine probe, which is scattered inflexibly, on striking the specimen surface. Several signals are generated from the sample as a result of the scattered electrons that can be collected and hence amplified to gather the associated information. An image is produced by scanning the specimen surface crossways using the probe and the generated image can be viewed on a digitized television raster. The multiple gathered gesture can be displayed on a screen whose time base is similar to that of the probe scan. The most usually utilized gesture for analysis is that from the subsequent electrons. The other signals that have been employed in order to acquire the

Fig. 3.2: Top: 28 kV SEM image pair of a 0.5% SWCNT-epoxy sample tilted at −5° (left) and at −5° (right) angles, 10.6 µm HFV. Bottom: 3D SEM image generated with these images. The red arrows point to a CNT very close to the surface and the blue ones to a CNT embedded at an estimated depth of 260 nm [7].

microstructural information are the typical X-rays, visible cathodoluminescence, scattered back electrons, and the remaining sample current. The performance of the scanning electron microscopes has been enhanced by the incorporation of advanced field emission guns [2].

The incident high-energy electron ray is focused by an electromagnetic probe lens to a diameter that might be limited to 2 nm. The current associated with the probe decreases with the reduction in the probe diameter. Much larger probe diameters are required by some signals to obtain the microstructural information, as is the case with X-rays emanating from the individual chemical constituents. In such cases, the specimen is energized with an incident ray that has low efficacy, owing to the smaller excitation crosssectional area. The size of the energized volume that lies below the surface of the sample also limits the firmness of the image obtained via X-rays, in addition to the statistics associated with the photon counting.

The characteristic of the X-ray signal aids in the quantitative determination of the associated microchemistry. This may used to indicate the presence of multiple constituting elements in the definite area of interest. X-rays are typically created in the sample volume that is situated below the probe and the diameter of probe is nearly 1 micrometer. This diameter corresponds to the depth of the electron infiltration into the specimen. There are three distinct formats in which the X-ray signals can be viewed: X-ray line-scan, X-ray spectrum, and X-ray data. In the case of X-ray spectrum, the concentration of the signal emanating from the chosen area is expressed as a function of the energy or the wavelength related to the characteristic X-ray. In the case of X-ray line-scans, the concentration of the X-ray is expressed and shown as a function of the place of the probe. The X-ray information can be observed as a basic map wherein all the photons that arrive at a specified energy gap are shown and observed in the form of color-coded dots [2].

The backscattered electrons provided information associated with the mass bulk or the usual atomic number of the sample. The denser regions are reflected as brighter images. The subsequent electron gesture is, however, of the greatest value of the many generated gestures. There are dual major details behind this: the number of subsequent electrons is more than the electron current associated with the primary ray and the second reason is the 100% efficacy associated with the collection of the secondary electrons. The subsequent electron originates within the surface layers of specimen owing to the limited average distance traversed by the low-energy electrons in the specimen. The incident ray entering the specimen as well as the scattered back high-energy electrons are the source of the subsequent electrons. The resolution associated with the secondary electrons need not be degraded by the inelastic dispersion of the primary ray. Moreover, there are no significant limitations in the resolution due to the high-energy electron flux. As such, the ability to focus the probe ray is one of the determining factors for the ultimate resolution of the image generated by the secondary electrons, and is typically of the order of 2 nm [2].

The preparation of the specimen for scanning electron microscopy is not complex. It is, however, to be borne in mind that when the specimen is less stable such as the polymeric samples, biotic tissues may be tarnished by the high-energy electron ray and therefore may contaminate the microscopic column as well as the specimen. A conductive coating is therefore employed for the nonconductive specimens in order to take care of the electrostatic charging. Another alternative is the use of low ray voltages. The contrast obtained in the image may be associated with the surface topography as well as the variations in the atomic number or the mass bulk of the specimen.

Stereoscopic analysis has not been explored to its full extent. A stereo image can be generated by slanting the specimen about a recognized axis and the images are then recorded at dissimilar slant angles. The stereo image can be used to observe the dispersal of the depth in the microstructural characteristics. The accurate measurements associated with the lateral displacements of the characteristics in the two different constituents of the observed stereo image can be achieved through the parallax measurements. The surface of the rough specimen can be observed in 3D in the scanning electron microscope through the process of stereo imaging.

The diffraction information is also contained in the backscattered electrons. A charge-coupled device aids in the extraction of information associated with the crystallography of the sample. Moreover, a suitable computer software package can be employed to automatically interpret the electron backscatter diffraction pattern. The orientations of the grain in the case of the polished and polycrystalline sample can be observed clearly in the color coding and displayed in a mode of operation, termed as orientation imaging microscope.

Chapter 4
Scanning probe microscopy

4.1 Introduction

Through the previously discussed microscopy techniques as such optical electron microscopy and transmission electron microscopy, it was observed that the there is an inverse relation between the source and the image planes of the sample under investigation. This is owing to the fact that deflection design resulting from the sample produced in the reciprocal place is concentrated in the image plane. On the other hand, an image that is produced from a sample is focused on the image plane, which is intricate by the angular dispersal of the probe radiation from emanating source.

The scientists at the National Bureau of Standards observed changes associated with the field emission current when a tungsten field emission tip is placed within close proximity of the surface of the sample. The changes related to the topology of the specimen surface were recorded when the sample surface was scanned perpendicular to the tip of the field of the emitter. This was the case 40 years ago and at that time it was not possible to maintain the dimensional constancy of the sample. Maintaining the dimensional stability was only possible with the advent of dependable piezoelectric drives. With the employability of such drives it became possible to not only maintain the mechanical stability but also the thermal stability. In the present scenario, a number of experimental methods are widely available to ensure that the studies can be made of the nanometer-sized interaction forces among the two solid objects. In addition, investigations related to the electrical possessions as well as atomic structures of the small areas can also be made.

All such techniques can be placed below the single banner of scanning probe microscopy. There are two instruments and in particular, scanning probe microscopes: atomic force microscope and the scanning tunneling microscope. The present chapter begins with the basics on surface force measurements and then terminates with the concluding remarks on the discussions made.

4.1.1 Surface forces

Between two solid surfaces, there are two components associated with interactions, that is, repulsive and attractive. The forces arising as a result may be either short-range or long-range. The presence of surface absorbates strongly affects the communication among the surfaces. The interaction is also exaggerated significantly by the existence of liquid or gaseous atmosphere between the two interacting surfaces.

An effortless two-body prototypical associated with the interatomic bonding between the atoms might be developed in case the long-range forces and the short-

https://doi.org/10.1515/9783110997590-004

range forces between atoms are considered to be attractive and repulsive, respectively. The developed model provides the information on the basis of the physical and chemical properties for the qualitative force of attraction and repulsion among the atoms. The identified physical and chemical possessions should be the heat of development, the tensile modulus of the material, interatomic separation between two solids, and the coefficient of thermal expansion. The model, when applied to the solid surfaces that are taken in the near presence will have to take into consideration the multi-body exchanges associated with atoms of the two surfaces. The effects associated with the local curvature as well as the surface roughness also have to be considered in the model application. As such, a three-phase model results that includes two solid phases, their interphases, and the environmental region. It is here again that consideration must be given to the long-range as well as the short-range forces. The long-range forces are experienced only at larger separations. Such forces are attractive and these larger separation forces are considered as noncontact regions and can be demarcated as expanding outside the variation point on the potential span diagram. This point of inflection resembles the extreme in the remaining attractive force. A contact region can be defined at a distance inferior to separation and corresponds to the potential energy that is equivalent to nil. The attractive forces within this regime are insignificant and hence there is dominance of the repulsive forces between two interacting bodies. A semi-contact region, on the other hand can be defined in between the contact and the noncontact regions. This zone encompasses the equilibrium spacing between the two surfaces and is positioned on both sides of the point where the potential energy reaches its minimum.

The connections in the three aforementioned areas can be seen detached in the atomic force microscope. The information can be used to gather the data associated with the behavior and dispersal of the forces. Coulombic electrostatic forces are the sturdiest long-range forces among two solids and this can be either attractive or repulsive in nature. The nature of the forces depends on the sign associated with the entire charge supported by the two surfaces. On the other hand, van der Walls forces dominate at shorter distances. The aforementioned polarization can appropriately be classified under three distinct categories. The permanent molecular dipole moments results in the strongest forces. Such forces are responsible for the creation of localized electric fields and as a result establish well-organized filling of an assemblage of surface dipoles. The neighboring atoms will also be attracted towards the electric field created by the molecular dipole themselves. The created attraction depends on the proportion of the strength of the molecular dipole as well as the polarity of the atoms cooperating with one another. This second type of polarized force is recognized as Debye interface. Lastly, it is the random oscillations in the electric fields of any polarizable atom that will give rise to confined and attractive exchanges with the polarized neighboring atoms. These will lead to the generation of forces that are referred to as dispersion or London forces.

There is a rapid decay of the van der Walls forces after they are created. The values associated with the dipole moment of the atomic or molecular objects are one of determining parameters for the strength associated with the interactions. Polarizability of the atomic or molecular entities is another factor determining the strength of interactions. Asymmetric molecular absorbates as such water vapor and carbon dioxide possess strong van der Waals adsorption owing to the associate large dipole moments. On the other hand, the symmetrical molecular assemblies, as for instance, methane, possess appreciable polarizability but insignificant dipole moment. Polar groups containing hydrogen, especially OH^- and NH_2^-, establish strong hydrogen bonds and are therefore sturdily attractive at very small distances.

The overlapping inner electron shells associate with themselves the repulsive forces that are strong and short-range. Such forces dominate when the solid surfaces are enforced jointly. The repellent also enforces the rule in the case of solid additives added into a liquid medium. The example that exhibits such characteristics is that associated with the inclusion of solid surfaces in a polymer matrix wherein the solid surfaces avoid coming into contact with one another due to an effect referred to as steric hindrance. Strong but unpredictable electrostatic forces are generated owing to the presence of charged surfaces in nonpolar, liquid, and gaseous conditions. A coat of counter-charge next to the charged solid surface is generated owing to the existence of polar surrounding that encompasses high dielectric constant. This coat of counter-charge is referred to as the Debye or double layer. There is great practical importance associated with the surface charge along with the disperse coat of the counter-charge. Such samples are readily investigated with the different available surface force measurement techniques.

One of the most fruitful theories that have been established to foresee the surface forces playing a critical role in the stabilization of colloidal dispersions is that established independently by Derjaguin and Landau in Russia [8] and Overbeek and Verwey in Holland [9]. The developed theory is known as the DLVO theory and has been improved with time and the modifications have been able to clarify the issues that identify the forces among two solid surfaces in the existence of liquid solutions, absorbate coats, and surface segregation.

It has been established earlier that the measurement of the surfaces forces that are usual on the surface of the specimen can be accomplished precisely. The measurement can be performed with the purpose of parting of two solid surfaces and with the advancements, the sub-nanometer upright resolution has been attained. Therefore, in order to explore the morphology of the solid surface, the consideration of the lateral resolution is a primary consideration when using a solid probe. This is also required to study the chemical and the physical properties of the specimen.

Flake of mica has a smooth surface atomically and is one of the featureless substrates. However, there is existence of bodily adsorbed moisture from surrounding conditions. There is insignificant interference of the cleavage steps on the surface of mica on the measurement of the associated forces. No noticeable surface is visible generally on such a featureless substrate. However, there are other materials the surface of which have certain morphological features that are associated with varying degrees of physical order and therefore are of significant applied importance. Surface roughness associated with the specimen is seldom random in terms of either the wavelength or the amplitude. Surface roughness may arise due to the scratches from the polishing operation or from the grinding process. The other prominent sources of surface roughness are chemical etching, corrosive pitting, etc. The surface roughness of the specimen that arises owing to anyone of the aforementioned processes depends greatly on the crystallographic structure. In case of polycrystalline solids, the equilibrium surface consists of grooved grain boundaries and facetted crystal surfaces. There may also be the presence of a second phase in such solids. The second phase may be present in the microstructure of the solid material or may arise owing to the deposition on the surface as covering or from erosion response. There may be situations wherein this second phase may be present in the form of a continuous slim layer or as intermittent, inaccessible, or interrelated islands.

The scanning probe will have an operative tip radius ranging from 2–200 nm in case the tip is a solid needle and the geometrical design can be approached by a pyramid or polished cone. The radius range is much greater in comparison to the interatomic spacing on any solid specimen surface. Examination of probe in scanning electron microscope can aid in determination of the probe tip radius. The estimated surface area of contact between the area of contact of the sample surface and tip of the probe is the smallest at 100 nm^2. The probe tip radius plays a crucial role in determination of the resolution of the scanning probe image. One of the main factors behind this phenomenon is the periodic variation in the number of atoms beneath the scanning probe as it moves across the surface, which directly influences the overall measured signal. With increasing tip radius, there will be a decrease in the comparative fullness of the differences in the force swaying. However, the wavelength associated with the force oscillations correspond to that of the atomic structure. These oscillations will be untraceable in cases where the tip radii of the probe are large. However, in case the tip radii of the probe is of the range of 10 nm then it becomes probable at times to resolve the typical interatomic spacing. This is irrespective of whether the calculated force is due to the numerous body interactions and is therefore measured on a large scale. However, there is often an argument that this is not the true 3D firmness and that it is only possible to resolve the surface of the atomic structure through the electrical measurements.

4.2 Scanning probe microscopes

In case of all the scanning devices, the info is collected pixel by pixel and is not combined over the period for which image is viewed. Hence there is wide difference between such sequential collection of the data in the scanning microscope and the data collection using the optical or transmission electron microscopes. In cases of transmission or optical electron microscopes the data is collected in parallel. There is very low rate of data collection in case of scanning probe microscope and the rate can be accelerated through the thermal and mechanical drift associated with both the specimen and the imaging setup. This also ensures that dependable info can be composed from solitary and inadequate quantity of pixels. Images are limited to 256 × 256 pixels and it is impractical that the info set associated with every of the image may surpass 10^5 pixels.

In the ensuing discussion, the chronology associated with the history of the different scanning probes employed has been presented. In 1936, the field emission of electrons was illustrated wherein a high vacuum was provided with a negative potential, the high vacuum being applied to the sharp tip of the refractory and flattened metallic cable. Spot welding was used to rigidly mount the needle shaped sample by spot welding it to the heating strand. The heating element was fixed to the two electrodes. Equation (4.1) depicts the field emission current i, obeying the Fowler-Nordheim equation:

$$i = A \frac{(\mu/\varphi)^{\frac{1}{2}}}{(\mu + \varphi)} F^2 \exp\left(-B \frac{\varphi^{1/2}}{F}\right) \qquad (4.1)$$

where Fermi energy is represented by μ, the electric field strength is denoted by F, ϕ is the work function, and A and B are constants. The projected image obtained due to the distribution of the field emission current over the hemispherical tip was revealed on the crystal assembly of the tip.

In 1956, the atomic resolution was primarily illustrated in the field-ion microscope. The image in case of the atomic resolution was achieved as a result of the field ionization of the gas ions taking place at the surface of the cryogenic chilling and positively charged metallic tip. A projection image of the crystalline metallic tip surface was then produced. In the projected image it was revealed that the atoms of the ends of the associated regularly filled atoms had sufficient resolution. It was also likely to ion-charge the atoms at the ends by increasing the power of the electric field at the tip surface. As a result, the ionized atoms could be removed from the surface through the use of field evaporation. This then aids in achieving a symmetric as well as smooth surface. Cryogenic cooling aided in cooling the electrodes of the sharp tip at cryogenic temperatures. The primary device, which was capable of attaining atomic firmness was the field emission microscope.

Mass-resolved chemical study of the specimens was made possible in the year 1967. This was achieved by carrying out the analysis on the field-evaporated ions from the surface of specimen in the atom probe through the use of time-of-flight mass spectroscopy. The achieved chemical analysis was at the sub-nanometer resolutions. The selected regions over the hole were brought on the fluorescent image screen by aligning the metallic tip. This was done to ensure the passage of the field-evaporated atoms into the mass spectrometer. Over the next 40 years, it was possible to detect the field-evaporated atoms over the broader angle and as result massive datasets were produced. These data sets encompassed information associated with the mass-resolved chemical study of the atoms field evaporated from the specimen needle.

The possibility of the scanning tunneling microscopy was illustrated in the year 1982. This was done soon after the thermal as well as the mechanical stabilities of the piezoelectric control associated with both the lateral as well as the vertical displacements were adjudged. The electronic structure analysis was possible by the use of the scanning tunneling microscope. The instrument was capable of providing atomic resolution. The existence of several crystal structures as well as the thermodynamically steady rearrangement associated with the stuffing of the surface atoms was well revealed by the use of the tunneling microscope. The sharp tip was affixed onto the provision frame that was in turn fixed rigidly to the piezoelectric actuator. The piezoelectric actuator was both mechanically stiff as well as cylindrically stable. The presence of the piezoelectric actuator with aforementioned characteristics ensured control at the sub-nanometer level of the separation between the specimen surface and the probe tip. This also ensured the lateral positioning of the probe.

In the year 1986, the development of atomic force microscopy took place. This was achieved by affixing the solid and needle-like probe at the end of the cantilever encompassing flexibility. The shifts in the cantilever that resulted owing to the surface contact forces were monitored. The employed flexible cantilever beam was embedded with the inflexible, piezoelectric x-y-z actuator element and a split-beam laser detecting system was used to observe the movement of the cantilever. There has been advancement in the earlier versions of the atomic force microscope and therefore, it requires no vacuum for its operation. As such the latest atomic force microscopy can work under atmospheric conditions or in a measured liquid or gaseous atmosphere. The tip of the probe has the potential to investigate an extensive scope of magnetic, electrical, and mechanical possessions owing to the resonant vibration of the cantilever. It also allows the atomic force microscope to function in three different modes: noncontact, semi-contact, and contact modes. This has been made possible with the regulation of the parting of the probe from the sample surface. The different modes of operation aided in revealing the details associated with the dominant forces for an extensive scope of engineering materials under diverse environmental conditions.

4.2.1 Atomic force microscopy

The elementary components of atomic force microscopy or the scanning tunneling microscope include the microscope, scanner, detector, probe, anti-vibration table, controller, digital signal processor, electronics interface, and graphics presentation. The tabletop device is affixed on a base that is rigid and has higher damping capacity. This base ensures that the extraneous vibrations are eliminated from the system. The compartment is enclosed entirely in order to make it available for environmental control as well as for thermal stability. In the case of atomic force microscope, a high vacuum requirement is not essential. The important components of the encompassed probe assemblage contain the mount for the probe assembly as well as the piezoelectric site regulation setup that has precision at the sub-nanometer level. The sample stage is versatile, enabling optical alignment of the specimen relative to the probe and facilitating coarse adjustments to control the spacing between the probe and the specimen.

A split-beam laser setup monitors the mechanical deflection of the cantilever and the associated frequency and amplitude of the shaking of the probe cantilever. The system is enclosed within the microscope chamber. The split-beam laser setup is provided with an external control mechanism for the alignment of the mechanical recognition setup. The scanning of the probe as well as the parting between the probe and the specimen are controlled analogously. All the constraints that are conveyed by the graphics juncture are digitized.

The cylindrical piezoelectric controller system is considered to be the core of the atomic force microscope. Application of voltage across the piezoelectric tube results in movement in the z-direction. Consequent to the applied voltage, the expansion as well as the contraction of the tube takes place. The shift in the x-y plane is through the application of the electrical energy of opposite sign crossways the aslope sections of the tube. This results in the deflection of the tube in either of the $\pm x$ or $\pm y$ directions.

There are two additional critical apparatuses that play a crucial role in the efficient working of the atomic force microscope. These components are the detection system and the design associated with the cantilever of the probe tip. V-shaped cantilever probes are supple about the points of provision. However, the V-shaped cantilever probes possess high torsional rigidity. These cantilevers can scan the specimen in a direction that is along the axis of the cantilever probe. There are other cantilever probes; however such probes are less flexible in comparison to the V-shaped probes. However, such probes are subtle to the torsional shifts taking around an axis that is parallel to the beam of the cantilever probe. Therefore, such probes can either scan in a direction parallel to the cantilever or in a direction that is perpendicular to the cantilever beam probe axis. While scanning in the perpendicular direction, the information associated with local frictional force acting parallel to the sample surface is provided.

There is wide availability of tip materials that have been used for successful functioning of the atomic force microscopy. However, silicon nitride is the most preferred of all the tip materials. This is because of the good resistance with regard to the tip damage owing to the better physical and chemical properties. The hysteresis losses as well as the associated mechanical phase delay in the generated signal are the two major factors that can be avoided by using the silicon nitride tip material. Hence, silicon nitride is considered to be the preferred material as far as contact mode imaging is concerned. However, owing to the tip radii possibly ranging from 20–60 nm, the atomic firmness required is outside the scope of the silicon nitride tips.

There are a number of other silicon tips that have been unified into a single-crystal silicon cantilever. The preparation of such tips is done through the process of focused ion beam milling. Such tips allow for higher resolutions to be achieved owing to the smaller tip radii, which is usually less than 10 nm. Hence silicon tips are mostly chosen for application in the semi-contact region. Such exploration can be achieved using a vibrating cantilever.

One of the critical components of atomic force microscopy is the recognition system that helps determine the displacement of the tip as it is scanned over the specimen surface. The detection system is based typically on the split photodiode finder system that operates with a solid-state laser source. On the deflection of the cantilever probe, the laser beam is reflected that scans crossways a minor opening among two halves of the photodiodes. As such, the signals produced hinge on the percentage of the light dropping on each half of the photodiode. In order to maximize the reflectivity, the cantilever is gold-covered and tuning of the finder system is accomplished by regulating the slant of the mirror. The maximum amplitude of the deflection is controlled by the span of the optical arm. A comparison between alternating current from the split photodiode and the drive signal can be made in case the cantilever is allowed to vibrate closer to the resonant frequency. Such a comparison can aid in the measurement of phase changes in the gestures that are associated with the nanostructural features on the surface of the specimen.

Application of the bias voltage is also possible to the cantilever probe and then the viewing can be done for the electric current among the sample and probe. This is, of course the base for the scanning tunneling microscope. However, some additional flexibility can be lent to the working of the atomic force microscopes.

4.2.2 Scanning tunneling microscopy

In case of scanning tunneling microscopy, the probe tip is scanned over the specimen surface, which is located in a vacuum chamber. This is done as a controlled separation between the probe and the specimen and the current-voltage characteristic that is identified at each of the pixel point. The Fowler-Nordheim equation forms the basis for the interpretation of the results. The local density associated with electron states at the surface of the sample can be probed accurately and the resolution of the individual surface atoms can be achieved owing to the sensitivity of the tunneling current to separating distance between the probe and the surface. Scanning tunneling microscopy is however restricted to the materials that are electrically conducting or semi-conducting. Maintenance of high vacuum inside the chamber is of utmost priority to prevent adsorption and adulteration of the specimen surface from the gaseous phase. The performance of the tunneling microscope depends on the thermal and mechanical firmness of the scanning probe setup and there are all possible chances of the drift taking place, so that the dimensions cannot be established for lengthy period of time. As such the number of pixels are limited from where the information can be collected for any specified image field. The entire number of pixels in a scanning probe is permanently lesser than that can be comprehended in a subsequent electron image.

However, inherent resolution of the scanning probe microscope is much better in contrast to the atomic force microscopy. The modes of operation are more restricted in scanning probe microscopy in contrast to the atomic force microscopy. This is owing to the detection of only the primary variations associated with the function as well as the surface density. It is worth noting that the function is not bulk possession and is surface possession. In particular, the function cannot be defined as a property that requires knocking out the electrons from the upper of the conduction band and taking it to eternity, rather it can be defined as possession that is suggestive of the energy required to just take the electrons outside the solid surface. Hence, the role is limited to the atomic packing of the atoms present at the surface associated to the place of extraction of the electron and can therefore be affected dramatically by surface contamination, crystal structure, and the presence of segregants and absorbates.

The current passing through the electrode tip is a function of work as well as the operative span between the tip of electrode and the specimen. The exponential variation of the aforementioned variations has been observed to be overlaid as the tip navigates the surface. As such, it is possible to operate the scanning probe microscope in more than one mode. The scanner assemblage is preserved at a persistent height as the probe is scanned across the x-y surface and the changes associated with the tunneling current are observed. The observed changes in current are then interpreted as either of the associated variables, that is, the work function or the probe-sample separation. On the other hand, it may be possible to select a given tunneling current and then adjust the height of the tunneling microscope accordingly. As such a feedback loop is created wherein the electron field current is kept at persistent levels as the tip scans over the surface. The variations associated with the tallness of the sample are then inferred as the variations in the specimen topography. Yet another alternative is to scan the tip for partial voltage at each pixel on the x-y surface and therefore get a value for the electrical energy requirement of the associated tunneling current. The slope of the curve associated with the $i(V)$ curve can then be understood in terms of the localized density of the states. Excellent atomic resolutions of the atomic packing as well as the surface structure are possible owing to the significant requirement of the local density states onto the atomic structure.

4.2.3 Field-ion microscopy

The advent of field-ion microscope took place four decades ago; however the results obtained during its operation were very limited on account of the material characterization. There were two primary reasons that could be attributed to the limited impact: one, the data sets collected contained only few counts as only a few thousand atoms were involved. As such the data collected was inadequate to offer sufficient numerical proof that they signified the actual chemical composition of the constituting basics within the specimen. The second reason was the necessary requirements for the tip of the sample needle, which was not only required to be metallic but also had to endure gigantic stress being used by the pragmatic electric field during the field evaporation.

The technology of atom probe was made possible with the advent of period-of-light mass spectrometry possessing sub-nanosecond firmness. The mass spectrometer was dependent on the time measurement associated with the ion to get to the indicator located approximately at a distance of 1 m in a long flight tube once evaporated from the surface of the tip on the application of voltage pulse. As such there was no dependence on the magnetic field application to analyze the mass spectrum associated with the ray of evaporated ions. The time required by the ion to reach the detector varied proportionately to the quantity of electric charge supported by the ion, the electronic charge, isotopic mass, and the charge on the ion.

It was therefore observed that the pure elements were able to provide the mass spectra consisting of several spectral lines. This was possible owing to the well-separated dissimilar isotopes of a solo element obtained by the mass spectrometer. Also, the advent of period-of-flight mass spectrometer resulted in the generation of ions with many charges from a solo atomic type. There are many situations wherein the molecular ions are composed from and are often related with the adsorbed remaining gases. The residual gases as such carbon dioxide and water vapor would have been stuck at the surface of the specimen.

The flight tube of the mass spectrometer has been shortened owing to the improvements made in the electronic detection. On the other hand, the collection angle associated with the tip has increased with the growth of channel-plate charge multipliers. The increased charge collection angle has in turn aided in the generation of the field evaporation image. The obtained data for every pixel associated with the CCD gatherer are accumulated. Moreover, there is no requirement to apply the voltage at the tip in order to excite the field evaporation. The energy barrier can be reduced with the application of the electric field. The energy barrier is associated with the field evaporation of the atom at any prearranged level. The pulsated laser can then be employed to activate the discharge of the ion from surface of the tip. As a result, there has been significant improvement in the period firmness of the mass spectrometer. There is the possibility that successive layer of particles can be mass-evaporated and the associated charge-mass ratio of the arrested ion can be noted as a role of both pixel site on the CCD indicator and also as a function of pulse number associated with the sequence of laser.

The period-off flight mass spectrometer has associated period-scale that relies on the flight tube distance as well as accelerating voltage. However, the time-scale is usually of the range of 100 ns. As such, period resolution superior to 100 ns is vital to ensure a decent top parting. Moreover, gigahertz electronics is essential to procedure the obtained consequences. The other important factors playing a critical role in the determination of mass resolution are the capacitance associated with the specimen tip and the extraction electrode.

If sufficiently sharp and distinct peaks are to be separated with unambiguity then there is need to recognize the charge on the ion as well as the isotopic mass before the identification for the chemical species is made. The applied electric field aids in determination of the charge on the ion field-evaporated from the tip of the sample. It is known that the ionic charge will be equivalent to the chemical valence of the element. However, the aforementioned knowhow is inappropriate. All the field-evaporated types are charged positively owing to the application of positive potential. The incorrectness can also be attributed to the fact that the energy that is required to field-evaporate an ion is identical to the ionization potential minus the product of the associated work role of the sample surface and the charge on the evaporated ion. The comparative values of the ionization potential required to excite an atom is a determining factor for the field strength required to evaporate the given charge. In most of the cases, the field-evaporated ions are charged twice and there may be cases when the ions are individually charged ions that are often found. There may be cases wherein the ions might be triple-charged. Though, it is the doubly charged ions that can rule, it might be probable for any component with greater charge to appear in the mass spectra.

A wide variety of substrates such as SiC, Pd, h-BN, Cu, etc. have been characterized using one of the scanning probe microscopy techniques known as scanning tunneling microscopy. Different degrees of corrugation have been revealed and the same have been depicted in Fig. 4.1. Scanning tunneling microscopy also provides information associated with the density of states for any 2D samples. This is accomplished through the combination of scanning tunneling spectroscopy and measurements done using the differential conductance. The measurements associated with the Dirac points of graphene have been accomplished through scanning tunneling spectroscopy. This has led to valuable information such as the correlation between the surface defects, grain boundaries, and the electronic properties of the single-layer crystals under investigation.

The energy position of the Dirac point as a function of applied gate voltage has been depicted in Fig. 4.2. The graphene sample has been strained through the tip of the scanning tunneling microscope and this has been done in the form of a Gaussian bump. This has been also employed to map the imbalance associated with the localized bulk of states. This aided in demonstrating the dominance of the pseudo-magnetic field to result in pseudospin polarization.

Another emerging field of research is the application of 2D materials for thermoelectric and thermal management applications. The reduction in lattice thermal conductance can be reduced strongly with the suitable bandgap engineering of graphene. The electronic conductance can also be maintained while enhancing the Seebeck coefficient. Scanning tunneling microscope technique has been employed to carry out atomic-scale mapping of thermopower for graphene. This characterization revealed the direct relationship of spatial distribution of thermovoltage with the electronic density.

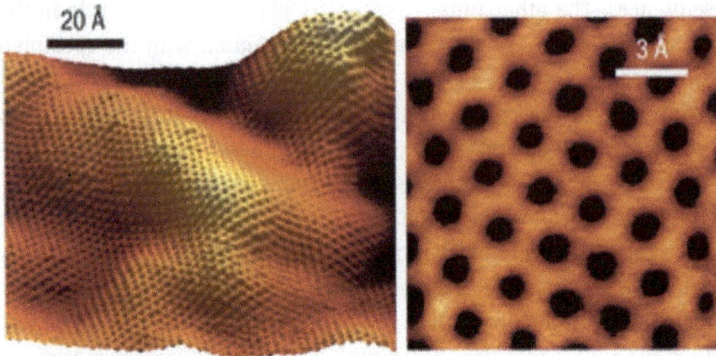

Fig. 4.1: Images from scanning probe microscopy [10].

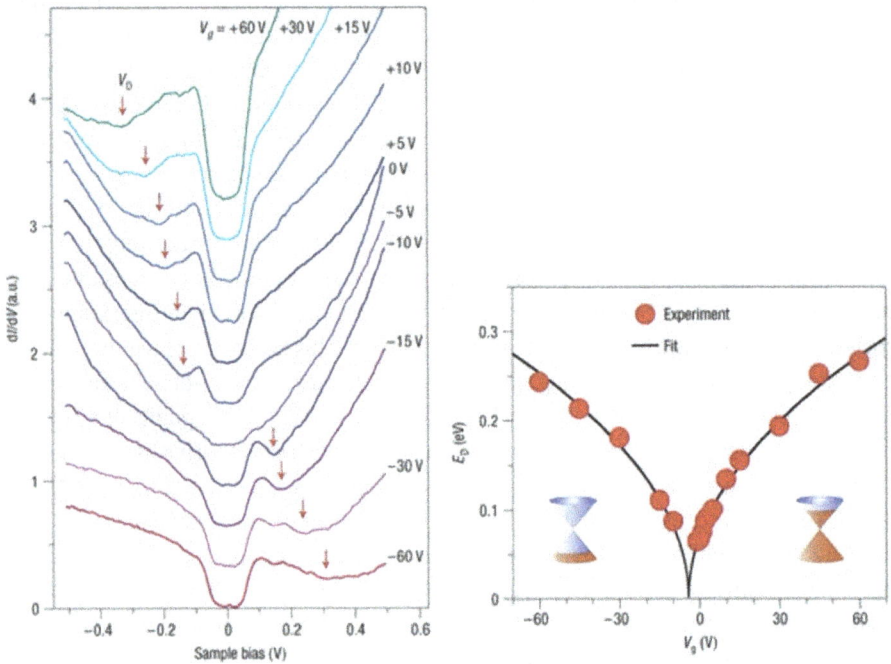

Fig. 4.2: The energy position of the Dirac point as a function of applied gate voltage [10].

4.3 Conclusion

There has been development of a few microscopes wherein the samples are the source of signals and the first such instrument that yields atomic resolution is the field-ion microscope. In the field-ion microscope, the sample is the tip of the sharp needle. The sharp needles are being employed as the source of electrons in the field emission cannon. As such high resolutions are obtained in case of scanning and the transmission electron microscopy. A similar needle shape is employed in the case of scanning tunneling microscopy as well as atomic force microscopy. The sharp probe in the case of the atomic force microscope is affixed on a cantilever beam and the its deflection is observed when vibrations are produced whereas the probe is scanned over the specimen surface.

Chapter 5
X-ray diffraction analysis

5.1 Introduction

When radiation interacts with a surface, it can undergo absorption or scattering. In the case of scattering, the process is elastic, meaning there is no change in the wavelength of the dispersed radiation and there is no energy loss. When a short-wavelength radiation interacts with the even arrangement of atoms within a crystal lattice, it leads to the formation of a diffraction pattern. The intensity of the dispersed radiation in the resulting diffraction spectrum is emitted away from the incident beam. This dispersed radiation intensity is then plotted against the scattering angle, which is two times the diffraction angle. The diffraction angles and the intensities of the diffracted rays are both influenced by the crystal lattice's structure. The Bravais lattice points and the dimensions of the unit cell significantly affect the diffraction angles, while the atomic properties of the atoms within the crystal structure show a central role in determining the intensities of the diffracted beams.

The analysis of the diffraction patterns or spectra can be approached from two perspectives. Firstly, in order to identify a crystalline material, a comparison is made between the diffracted angles, peak positions, and the relative intensities observed in the spectrum, with a known standard diffraction pattern. The diffraction spectrum shows a central role in identifying the crystal structure and determining the specific crystalline phase. The second approach involves comparing the acquired diffraction spectrum with a measured spectrum from a theoretical model that reflects the crystal structure. This method allows for further analysis and confirmation of the crystal structure. The confidence in considering the selected model as equivalent to the actual crystal structure depends on the degree of agreement between the predicted spectrum and the experimental data.

The typical approach includes a comparison of the calculated spectrum with the reference data. However, if notable inconsistencies emerge between the standard spectra and the measured data, it becomes essential to explore alternative models that can better explain the observed results. The progress in computer-based methods have greatly enhanced the flexibility and interpretability of crystallographic data. These advancements have mitigated the uncertainties and intricacies associated with traditional procedures, leading to more robust and reliable analyses.

5.1.1 Radiation scattering by crystals

For a crystal-like material to display a noticeable diffraction design, the wavelength of the radiation employed must be compatible with the interatomic spacing within the

https://doi.org/10.1515/9783110997590-005

crystal structure. This ensures a constructive interference within the crystal lattice. Different sources such as X-rays, neutrons, or high energy electrons can be employed to gather structural information about the lattice. While X-ray and high-energy electron sources are widely accessible, the availability of neutron sources remains a notable challenge.

The choice of radiation for obtaining a diffraction pattern determines the required dimensions of the specimen under analysis. Neutron-transparent materials necessitate specimens that are several centimeters thick, while X-rays are absorbed more strongly by engineering materials, limiting diffraction data to sub-millimeter surface layers, small crystals, and fine powders. In transmission electron microscopy, electrons are utilized, including high-energy electrons, with voltages limited to about hundred kilovolts (kV). At such high energies, inflexible scattering becomes dominant once the sample depth surpasses one-tenth of a micrometer, restricting the electron diffraction data to sub-micrometer sample depth. As a result, the diffraction data obtained from the same crystal using X-rays, electrons, and neutrons can differ based on the sample volume, which has significant inferences for sample dimension, facts reading, and specimen selection and preparation procedures.

5.1.2 Laue equations and Bragg's law

The collection of atoms interacts by the corresponding incident ray (incident at an angle 00), results in the scattering of a parallel ray of radiations at an angle with the wavelength. This scattering process creates a pathway alteration (A) between the dispersed and the incident beam. The pathway alteration can be calculated using the following equation:

$$\Delta = (y - x) = a(\cos \alpha - \cos \alpha_0) \tag{5.1}$$

where the interatomic spacing is denoted by a. The scattered and the incident beam will be in phase in case h is an integer. Let us consider a crystal lattice that is constructed of a three-dimensional array of atoms and is represented by lattice points that are regularly spaced. These lattice points are located at the corners of the primitive unit cell that has the lattice parameters a, b, and c. The following three set of equations, referred to as Laue equations, depict the condition for the scattered beam to be in phase with the incident beam [2].

The requirement for the dispersed ray to be in sync with the incident ray can be represented as $\Delta = h\lambda$, where "Δ" denotes the pathway alteration, "h" is a number, and "λ" stands for the wavelength of the radiation. Let us study a crystal-like lattice composed of a regularly spaced 3-dimensional arrangement of atoms. The lattice points are situated at the bends of the primitive unit cell, which is characterized by the lattice constraints "a", "b", and "c". The Laue equations, consisting of three sets of equations, describe the conditions under which the dispersed beam is in sync with the incident ray:

$$\Delta = a(\cos \alpha - \cos \alpha_0) = h\lambda \tag{5.2a}$$

$$\Delta = b(\cos \beta - \cos \beta_0) = k\lambda \tag{5.2b}$$

$$\Delta = c(\cos \gamma - \cos \gamma_0) = l\lambda \tag{5.2c}$$

The directions of the incident and the dispersed radiations in a crystal lattice are determined by the cosines angles α, β, and γ and α0, β0, and γ0. These directions are defined with regard to the unit cell of the crystal-like lattice. In crystallography, Miller indices (*hkl*) are used to represent specific crystallographic planes or directions within the crystal lattice. The Miller indices are determined by taking the complementary of the intercepts made by the plane or direction on the unit cell axes, and then reducing the resulting numbers to their smallest integer values.

Equation (5.3) represents a generalized and convenient form of the Bragg equation, which is a geometric relation used to determine the angular dispersal of the top intensities in a deflection spectrum:

$$n\lambda = 2d \sin \theta \tag{5.3}$$

The relationship described by eq. (5.3), known as the Bragg equation, establishes a relationship between the angular dispersal of the top intensities in a deflection spectrum. The equation incorporates key variables such as "*n*" (an integer), "*d*" (the space of the crystal plane), "*λ*" (the wavelength of the radiation), and "*θ*" (the angle of diffraction, measured between the incident beam and the lattice planes).

In accordance with the Bragg Law, the plane of atoms accountable for diffraction peaks are considered to act as specular mirrors. This assumption implies that the angle of incidence, O, is identical to angle of reflection. The term $(x - y)$ represents the path change between the incident beam and the beam reflected from two successive planes. The angle of scattering between the reflected and the incident beams is 2θ, $y = x(\cos^2\theta)$. By utilizing the identity $\cos(2\theta) = 1 - 2\sin^2\theta$ and recognizing that $x(\sin\theta) = 1$, the equation can be simplified to: $(x - y) = 2d \sin\theta$.

The interplanar spacing, represented by "*d*," can be described by utilizing the Miller indices of the crystal plane and the corresponding lattice constraints:

$$\frac{1}{d^2} = \frac{1}{V^2}(S_{11}h^2 + S_{22}k^2 + S_{33}l^2 + 2S_{12}hk + 2S_{23}kl + 2S_{31}lh) \tag{5.4}$$

The volume of the unit cell, denoted as *V*, can be determined using eq. (5.5):

$$V = abc\sqrt{1 - \cos^2\alpha - \cos^2\beta - \cos^2\gamma + 2\cos\alpha\cos\beta\cos\gamma} \tag{5.5}$$

The constants in eq. (5.4) can be obtained by the following set of equations:

$$S_{11} = b^2c^2\sin^2\alpha \tag{5.6.1}$$

$$S_{22} = c^2a^2\sin^2\beta \tag{5.6.2}$$

$$S_{33} = a^2b\sin^2\gamma \tag{5.6.3}$$

$$S_{12} = abc^2(\cos\alpha\cos\beta - \cos\gamma) \tag{5.6.4}$$

$$S_{23} = a^2bc(\cos\beta\cos\gamma - \cos\alpha) \tag{5.6.5}$$

$$S_{31} = ab^2c(\cos\gamma\cos\alpha - \cos\beta) \tag{5.6.6}$$

The aforementioned equations for an orthorhombic lattice reduces to the following:

$$\frac{1}{d^2} = \left(\frac{h}{a}\right)^2 + \left(\frac{k}{b}\right)^2 + \left(\frac{l}{c}\right)^2 \tag{5.7}$$

Bragg's equation, expressed as $n\lambda = 2d\sin\theta$, incorporates an integer value denoted as n, which signifies the order of reflection. When $n = 1$, it corresponds to the first-order reflection and represents the pathway alteration between the incident and the reflected beams from the (hkl) planes. Similarly, a second-order reflection agrees to a pathway alteration of two wavelengths. However, according to the Bragg's equation, a pathway alteration of two wavelengths is equivalent to a single wavelength pathway alteration for the planes of atoms situated at one-half the spacing, d_{hkl}. These planes possess the Miller indices, 2h, 2k, 2l. Therefore, it is customary to consider the nth-order reflection originating from planes with the Miller indices, nh nk nl, and spacing, d_{hkl}/n. The Bragg's equation for the nth-order reflection can be expressed as $\lambda = 2d_{hkl}\sin\theta$. For example, the first-, second-, and fourth-order reflections from the 110 planes can be denoted as d_{110}, d_{220}, and d_{440}, respectively [2].

5.1.3 Allowed and Forbidden wavelength reflections

In Bravais lattices that are body-centered or face-centered, specific planes composed of lattice points can exhibit destructive interference for certain orders of reflections. In the case of a body- centered cubic lattice, the reflections will be in phase for lattice points at coordinates l/2, l/2, l/2 for all orders of the 110 reflections. Conversely, destructive interference will occur for all the odd orders of 100 reflections, including 100, 300, 500, and so on. To determine the permissible reflections for body-centered and face-centered Bravais lattices, a helpful list can be constructed based on the integer value $(h^2 + k^2 + l^2)$. The Bragg's-equation can be reorganized for a cubic geometry as $(h^2 + k^2 + l^2) = 2a\sin\theta/\lambda$, resulting in a rectangular arrangement of the diffracted beams. However, certain integer values are absent from the list of permissible reflections, with 7 being an example. Additionally, certain integer values may correspond to multiple reflections, such as the 221 and 300 planes, which deflect at identical angles.

Forbidden reflections denote reflections that are not observed in a crystal lattice. To determine these forbidden reflections, specific guidelines have been established. In face-centered cubic (FCC) lattices, reflections are permitted when each of the Miller indices is either odd or even, while they are forbidden if the Miller indices consist of a combination of odd and even numbers. In the case of body-centered cubic (BCC) lattices, reflections are allowed when the sum of the indices is even, but forbidden when the sum is odd. These guidelines allow for the identification of the diffraction peak sequence. For instance, the presence of paired reflections such as Ill and 200, or 311 and 222, indicates FCC symmetry. Therefore, accurate computation is essential for determining the crystal's symmetry based on a specified diffraction pattern.

In certain cases, crystal lattice points can comprise clusters of atoms instead of single atoms, and the scattering intensity can differ based on the atomic number of the atoms involved. For example, in the face-centered cubic (FCC) NaCI structure, the lattice points are occupied by sodium cations, whereas the chlorine atoms are situated off lattice-points at the lattice vector ½, 0, 0. Two distinct lattice configurations exist: a mixed plane where the planes contain both anions and cations, and another configuration where there is an alternating arrangement of equidistant pure anions and cations.

In the mixed plane, the scattering of the two species occurs in phase due to the coplanarity of the anions. As a result, the intensity of the diffracted peaks is increased through the extra scattering from another species. On other hand, the alternating planes of anions a well as cations scatter in the opposite phase for every odd reflecting plane, causing a decrease in the intensity of the diffracted beams. However, for even-order reflections, the same plane of alternating anions and cations scatters in phase, resulting in an increase in the associated intensity. The extent of the reduction in intensities or their enhancement depend on the differences in atomic numbers and their respective scattering strengths.

5.1.4 Complementary space

Bragg's law states that the diffraction angle is in reverse relation to the space between the reflecting planes in the crystal lattice. Therefore, it is essential to define a 3-dimensional coordinate system where the axes are scaled in terms of inverse length. This coordinate system, known as complementary space, is established to facilitate the analysis of diffraction patterns.

5.1.5 The construction of limiting sphere

The range of values for l/d in Bragg's law is restricted to the interval between 0 and 2/λ, which corresponds to the permissible values of *sin θ* ranging from −1 to 1. This condition ensures that parallel planes of atoms can produce a diffracted beam. When the incident radiation is aligned with the x-axis and the crystal responsible for diffraction is situated at the origin, a spherical boundary with a radius of 2/λ is established. This boundary, known as the limiting sphere, encompasses all the valid points in the complementary spacing and defines a plane within the crystal that can diffract the incident beam of wavelength λ. Now, imagine a smaller sphere, referred to as the reflecting sphere, with a radius of l/λ. This sphere is positioned within the limiting sphere, touching the crystal surface and the limiting sphere along the x-axis. The line passing through this reflecting sphere, parallel to the diffracted beam, forms an angle of 2θ with the x-axis and intersects the surface of the reflecting sphere at a point denoted as P.

5.1.6 Bragg's law and its vector representation

Let us examine two vectors k_0 and k, which correspond to the wave vectors of the diffracted and the incident beams, respectively. We set the magnitudes of both the vectors as $|k_0| = |k| = l/\lambda$, where l represents the magnitude of the wave vector and λ denotes the wavelength. In this scenario, if the complementary lattice vector OP is equivalent to the vector g, then the magnitude of g is denoted as $|g| = l/d$. Here, d represents the crystal-like lattice spacing plane responsible for diffraction.

The equation $\lambda = 2dsin\theta$, known as the Braggs law, can be represented as $k_0 + g = k$ in vector form. In this equation, k_0 represents the wave vector of the diffracted beam, g represents the complementary lattice-vector, and k represents the wave vector of the incident beam.

5.1.7 The commentary lattice

In the complementary space, a lattice can be defined as is equivalent to the Bravais lattice in real space. The beginning of the coordinate system in the complementary space corresponds to the position of the crystal in the real space, specifically (000). This origin serves as the center for constructing the limiting sphere. Complementary lattice vectors, denoted as g_{hkl}, are constructed from the origin to the lattice points in the complementary space. These complementary lattice vectors are perpendicular to the reflecting planes with the Miller indices (hkl). The distance between the origin (000) and the lattice point hkl is equal to $1/d_{hkl}$, where d_{hkl} represents the spacing between the reflected planes. The sequence of equal distant complementary lattice points along a straight line represents the successive orders of reflection, as the lattice points lie on straight lines in the complementary space originating from (000). It is worth noting that in this representation, negative values of n (order of reflection) are valid and should be taken into consideration.

To ensure compliance with the Bragg's law, it is necessary for the complementary lattice vector to be contained within the reflecting sphere. This requirement is obtained by adjusting the wavelength λ or θ.

At this stage of the discussion, we will introduce the concept of complementary lattice unit cell, which is known by dimensions of the Bravais lattice unit cell as follows:

$$a^* = \frac{bc}{V}\sin \alpha; \quad b^* = \frac{ca}{V}\sin \beta; \quad c^* = \frac{ab}{V}\sin \gamma \tag{5.8}$$

The volume (V^*) of the complementary lattice unit cell is related to the unit cell of the Bravais lattice. It is worth noting that the axes of the complementary lattice, denoted as a^*, b^*, and c^* are not parallel to the axes of the Bravais lattice unit cell, represented by a, b, and c. This observation IS constant with the earlier discussion on directions and the Miller indices. Specifically, the direction indices [1001, [010], and [001] correspond to the axes of the unit cell in the complementary space, while the Miller indices (100), (010), and (001) represent the axes of the complementary unit cell. This distinction highlights that the complementary lattice axes are not parallel to the Bravais lattice axes' unit cell [2].

5.2 Methods for X-ray diffraction

Let us examine the reflecting and limiting spheres more closely, this time taking into account the superposition of the complementary lattice. There are three main factors that affect the ability of a crystal plane to generate diffraction peaks: the wavelength of the incident beam, the angle of the incident ray relative to the crystal structure, and the effectual dimension of the complementary lattice tip. As the X-ray wavelength decreases, the diameter of the limiting sphere increases. This larger diameter corresponds to the shorter complementary lattice-vectors, which in tum corresponds to the smaller interplanar space. When the crystal lattice is rotated around its center, the reflecting sphere sweeps through the available lattice points. Consequently, dissimilar diffracting planes follow the Bragg's law in a sequential manner.

When two axes are perpendicular to each other, all the lattice points within the limiting sphere have the potential for contributing to diffraction. The diffraction takes place at the Bragg angle (θ_{hkl}), where the diffraction condition is precisely met. In real-world applications, the finite size of the complementary lattice points is mainly calculated by the dimension and the quality of the crystal structure. The pathway length of the radiation is restricted by the absorption of energy from the incident beam, reaching a threshold where elastic scattering prevails over inelastic absorption.

As a result, the dimensions of the crystal structure that effectively donate to tenacious diffraction are limited.

The absorption coefficient plays a crucial role in determining the absorption effect, and it typically results in path lengths on the scale of tens of micrometers. In such cases, the beam tends to become blurred due to the prevalence of inelastic scattering. The effective diameter of the complementary lattice points is typically on the order of 10^{-5}, corresponding to the interplanar spacings of a few tenths of a nanometer. By accurately determining the lattice parameters of the crystalline structure with an error margin of approximately 10 ppm, it becomes feasible to assess the changes in crystal dimensions caused by factors such as temperature fluctuations, alloy additions, and applied stress. The measurement of the diffraction angle ranges for small crystallites and highly deformed crystals enables the observation of diffraction from specific planes with Miller indices hkl. This provides quantitative information regarding the crystal size and the degree of perfection. Overall, precise measurements and analysis yield valuable insights into the crystal size, perfection level, and the variations in dimensions under different conditions.

During our earlier discussions, we assumed a perfectly parallel and monochromatic incident beam. However, the consequences of this assumption becomes apparent when we consider the construction of the reflecting sphere.

If the incident beam exhibits a range of wavelengths, the ideal limiting sphere transforms into a shell shape. This results in the formation of a crescent-like region where the complementary lattice points satisfy the Bragg's law. Within this reflecting shell, diffraction peaks are generated only for specific wavelength ranges. Similarly, when the incident beam is not perfectly parallel, the reflecting sphere undergoes rotation around the center of the limiting sphere. This rotation corresponds to the divergence or convergence angle of the incident beam. As a result, two crescent-shaped volumes are formed where the Bragg law is fulfilled. These deviations from perfect parallelism and monochromaticity introduce faults in the identifying lattice space. The magnitude of the complementary lattice vector ($|g|$) and the angle between the incident beam and the crystal-like lattice are critical factors that contribute to these measurement uncertainties in the lattice parameters.

5.3 The X-ray diffractometer

An X-ray diffractometer includes vital components such as an X-ray source, a diffractometer assembly, an X-ray generator, a detector gathering, and systems for X-ray data cluster and dispensation. The diffractometer assembly plays a key role in controlling the alignment of the X- ray beam as well as the positioning of the X-ray detector and the sample The X-ray source generates the primary X-ray radiation, while the X-ray generator regulates its characteristics. The detector assembly captures the diffracted X-rays and converts them into measurable signals. The systems for X-ray data cluster and the dispensation handle the acquisition as well as the processing of the detector signals. Together, these components enable an accurate alignment, controlled positioning, and efficient data collection for X-ray diffraction experiments [2].

X-rays are produced via accelerated electrons towards a metal target housed within a vacuum tube. The high-energy electrons displace lower-energy electrons off their ground state, resulting in the creation of electron holes. As these holes are filled, X-rays are emitted. The energy of the electrons, measured in electron volts (eV), can be transformed into X-ray frequency, using the equation $eV = h\, \upsilon$, where h represents the Planck's constant. The wavelength of the X-rays, denoted as λ, can be determined by means of the equation $\lambda = c/\upsilon$, where c represents the speed of light. The upper limit of the X-ray frequency is determined by the condition that all the energy of the electrons is utilized to generate photons, while the lower limit is also set by the same condition. The minimum wavelength of X-rays is inversely proportional to the accelerating voltage and can be calculated using the relation $\lambda_{min} = 1.243/V$, where λ refers to the wavelength in nanometers and as V refers to the accelerating voltage in kilovolts.

Beyond the least wavelength, X-ray wavelengths exhibit a continuous distribution. This distribution is formed due to the intensity of the incident beam, which is directly influenced by the beam current and the electron energy. Additionally, the intensity of the incident beam is affected by the atomic number of the target material, with higher atomic numbers yielding higher intensities. The X-rays released from the specimen display a continuous range of photon energies and their corresponding wavelengths. This phenomenon is commonly stated to as white radiation or Brernsstrahlung.

The constant spectrum of white radiation, characteristic of the chemical constituents, is accompanied by a sequence of slim and concentrated peaks. These peaks are a result of specific energy releases when electrons from higher energy levels fill the electron holes created during collision events. When an electron is ejected from a lower energy shell, such as K-shell, the atom is excited to a higher energy state, E_K. The energy of the empty K-shell will decay to E_L if it is occupied by an electron from a lower energy shell, such as the L-shell. This energy difference, $E_K - E_L$, manifests as an X-ray photon with a specified wavelength, known as Kα line, in the characteristic goal spectrum. Alternatively, if the empty K-shell is occupied by an electron from an even lower energy shell, such as the M-shell, Kβ photons may result. These photons have a shorter wavelength compared to the Kα line in the characteristic spectrum. Consequently, the residual energy of the atom excited in the E_M state would be lower than that in the E_L state.

Based on the preceding discussion, we can deduce that the energy transition of the excited atom from the EL and EM energy states will lead to the emission of a considerably longer wavelengths. The typical L and M spectra entail an interstice line, as there are multiple possible sources for donor electrons to fill the holes in the L- or M-shells. If the energy of the incident electron exceeds the excitation energy of the atom, a characteristic line can be produced in the incident beam. With increasing atomic number of the goal material, the stimulated power vital to expel an electron from a specific inner shell also rises. This is attributed to the stronger binding of electrons to the nucleus in atoms with higher atomic numbers. Elements with lower atomic numbers, such as those in the initial row of the periodic table, will solely exhibit K-lines since their electrons primarily occupy the K-shell. In contrast, elements with higher atomic numbers will present M and N-lines in their spectra, resulting in an overall highly intricate spectra [2].

The intensity losses arising due to inelastic scattering needs to be avoided or minimized in case the elastic scattering of the X-rays dominates the interface with the sample. The rate of intensity loss for a unicolor X-ray beam crossing a thin sample in the x-direction is given by $dI/dx = -\mu I$, where μ is the linear absorption coefficient for X-rays The important consideration is the mass of the material that the X-rays traverses rather than the thickness of the sample. The intensity that is transmitted is then given by eq. (5.9) as follows:

$$\frac{I}{I_0} = exp\left(-\frac{\mu}{\rho}\rho x\right) \tag{5.9}$$

In most materials, the mass absorption coefficient generally exhibits a wavelength-dependent increase. However, this increase occurs in a series of distinct steps known as assimilation edges. These edges agree to specific wavelengths where the X-ray photons possess adequate energy to ionize atoms by removing the inner part of shell electrons in the specimen. These absorption edges represent the minimum excitation energies required for X-rays. Additionally, short wavelength, high-energy X-rays can induce the generation of secondary characteristic X-rays through fluorescence. These ancillary characteristic X-rays have lengthier wavelengths within the sample target and contribute to the phenomenon of fluorescence. In X-ray fluorescence excitation, the generation of white radiation is absent, resulting in a distinct absence of background.

In X-ray diffraction measurements, the selection of radiation with a wavelength near the absorption minimum is crucial to minimize absorption by the incident beam and avoid fluorescence. Consequently, using Cu Kα radiation ($\lambda = 0.154$ nm) is not ideal for measuring diffraction X-rays on steels as well as iron alloys. On the other hand, Co Kα radiation with a slightly longer wavelength of $\lambda = 0.1789$ nm is a better choice. It falls just beyond the lengthy wavelength side of the K_{Fe} assimilation edge, enabling observation of sharp diffraction patterns from the steel without interference from fluorescence.

The parameter μ/ρ plays a crucial role in determining the appropriate sample thickness for achieving the desired reduction of the incident beam intensity to I/e. When considering Co Kα radiation, the values of g/p for aluminum and iron are calculated to be 67.8 and 47, respectively. By incorporating the densities of aluminum (2.70 g/cm^3) and iron (7.88 g/cm^3), the corresponding sample thicknesses can be determined. For iron, the calculated thickness is 54.6 pm, while for aluminum it is 27.6 gm. These values represent the required sample thicknesses to obtain the desired diffraction signal from each metal when utilizing Co Ka radiation.

The experimentations associated with the X-ray diffractions require either white radiation or the monochromatic radiation. Excitation of the K-radiations from the pure metal target results in the generation of monochromatic radiation. This is accomplished by filtering the β component in the incident beam through the use of a foil that strongly absorbs this component and therefore there is no reduction in the associated α component in the incident beam. The chosen filter should have the absorption edge falling in between kα and kβ wavelengths. One of the good examples is the use of nickel sieve with the target metal being of copper, which transmits the Cu Kα beam but restricts the Kα beam.

Utilizing a single crystal monochromator is an effective method for selecting a monochromatic beam in X-ray diffraction. The monochromator crystal is oriented in a way that enables the incident beam to be diffracted at the typical Kα peak. The resulting diffracted unicolor beam can be utilized as a radiation source or employed as a filter for the diffracted signal. To ensure that the radiation from a line source satisfies Bragg's condition, the monochromator crystal is curved into a circular arc. This curvature enables the radiation from the line source to be diffracted by the monochromator crystal and directed toward a specific position on the specimen. Similarly, the detection system can be positioned in the track of the beam diffracted by the sample, with the monochromator serving as a filter for the detected signal.

The rotation of the X-ray detector plays a vital role in capturing the X-ray spectrum during diffraction experiments. The specimen is mounted on a goniometer stage within the diffractometer, enabling rotation of the sample around one or more axes. Precise alignment and calibration of the diffractometer are crucial for achieving an optimal resolution in the diffraction method, typically with a calibration precision of 0.01°. In particular, when utilizing a bent monochromator in a concentrating diffractometer, precise positioning of the specimen is recommended to compensate for any displacement of the sample plane, which can lead to a shift in the apparent Bragg angle. To ensure the perpendicularity of the specimen plane to the radius of the concentrating circle, the detector should be rotated round the focusing circle at a rate that is twofold that of the specimen rotation. These adjustments and alignments are essential to obtain accurate and reliable diffraction data [2].

The proportional counter is widely used as a detection system in X-ray diffraction experiments. It operates by ionizing a low-pressure gas when the incident photons interact with it, resulting in the formation of a cloud of charged ions. This ionized cloud is then collected by way of an existing pulse. The magnitude of the current pulse is directly proportional to the energy of the incident photon. To ensure that only the desired signal is captured and the stray photons are eliminated, electronic discrimination techniques can be employed. The current generation proportional counters exhibit excellent energy resolution, reaching values better than 150 eV. This high resolution enables effective suppression of the background noise associated with white radiation. However, it should be noted that proportional counters have a dead time associated with each current pulse. This means that if a second photon arrives within a microsecond of the first photon, it will not be detected. As a result, there is a higher limit on the count rate and the peak intensities noted at higher count rates may be undervalued due to this limitation.

The extreme thick length of a specimen that can be analyzed using X-ray deflection is limited by the mass assimilation factor of the incident beam, while adjacent sizes depend on the diffractometer's geometry. In automation of dust diffractometers with Bragg-Brentano configuration, the perpendicular breadth of the illuminated area is typically around 10 mm, and the span of the irradiated sample varies with the incident angle, typically ranging from 1 to 7 mm. The size and spacing of the Soller slits or divergence slits determine the collimated beam's area, which decreases with increasing diffraction angle in fixed slits but remains constant for sufficiently thick samples. Some Bragg-Brentano diffractometers have automated the divergence slits that widen the incident ray as the deflection angle rises, expanding the illuminated size. Consideration of the diffraction volume's impact is important when determining integrated intensities in X-ray diffraction experiments.

5.3.1 Powder diffractions

X-ray diffraction analysis in crystalline engineering materials is affected by the fact that the grain size is usually smaller than the material's thickness. This applies to both the dispersed and the compacted powder samples. The term "powder diffraction" encompasses both the characterization of the resulting diffraction pattern and the subsequent analysis required to interpret the diffraction data obtained from polycrystalline samples.

When considering randomly oriented grains that are smaller than the cross-section of the incident beam, the wave vector k_0 of the incident ray can be in any direction in the complementary plane. This means that the complementary lattice is free to rotate around its origin. For Bragg reflection to occur, the oriented grains must have g vectors that trace the surface wall of the reflecting sphere, resulting in the generation of diffraction cones. Each diffraction cone corresponds to specific lattice planes with different d-spacing. The innermost diffraction cone represents ice planes through the major d-spacing. The radius of reflecting sphere determines the minimum detectable d-spacing, which can be calculated as $d_{min} = \lambda/2$, where is the wavelength of the incident beam. By rotating the detector around the normal of the incident beam and recording the intensity of the diffracted beam as a function of the diffraction angle, a diffraction spectrum for the sample can be obtained.

By continuously rotating the specimen at a constant rate $(d\theta/dt)$ around an axis perpendicular to the diffraction plane and simultaneously rotating the detector at twice that rate $(d(2\theta)/dt)$ around the same axis, the normal to the diffracting planes in the crystal that contributes to the diffraction spectrum remains parallel. This technique is particularly useful for nonrandomly oriented polycrystalline samples. By applying processes like plastic deformation or directional solidification, the grains in the sample can be aligned along a specific direction. This alignment is commonly observed in thin-film electronic devices produced through chemical vapor deposition. Samples with such crystalline texture and preferential orientation may display enhanced or reduced diffracted intensities for specific crystallographic reflections, compared to randomly oriented polycrystals, depending on the coincidence between the normal to the diffracting planes and the direction of the bulk material [2].

Texture, also referred to as preferred orientations, plays a crucial role in numerous material applications due to the inherent anisotropy of the mechanical, physical, and chemical properties. The presence of preferred orientations holds significant influence over material behavior. For instance, in silicon iron, the magnetic hysteresis displays pronounced variations along <100> and <111> directions. Transformer steels are deliberately processed to exhibit a strong texture, leading to reduced hysteresis losses. However, certain mechanical engineering applications require a lack of surface texture. In such scenarios, structural steel undergoes cross-rolling to minimize grain alignment and the formation of preferred orientations. This approach ensures a more isotropic behavior of the material.

The lattice spacing in a polycrystalline material is directly influenced by the state of stress present in the sample. Stresses can originate from various factors, including operational conditions or improper assembly of components, leading to changes in the lattice spacing. For instance, tightening a nut can induce tensile stress in a bolt. Additionally, residual stresses may arise from the material's processing history, such as variations in plastic work or specialized surface treatments like ion implantation, shot peening, or chemical modifications. Irrespective of the cause, it is essential for the material to maintain mechanical equilibrium, even in the absence of external forces, to ensure internal stresses are balanced within the material.

Accurate determination of the residual stress in a material can be achieved by utilizing X-ray techniques to measure the lattice spacing. The lattice strains, which reflect the stress within individual grains, are measured and then converted into stress values using the elastic constants of the phases present. It is crucial to distinguish between two types of stresses: micro stresses and macro stresses. Macro stresses occur when large number of neighboring crystals experience the same stress, leading to a smooth stress distribution throughout the sample and an overall equivalent stress state. An example of macro stresses is the presence of compressive stresses on the surface, balanced by tensile stresses. On the other hand, micro stresses occur when stress significantly varies in magnitude and direction between individual grains, while the average stress in the component is zero. An example of micro stresses is the stress resulting from the anisotropy of thermal expansion in noncubic polycrystals, which imposes constraints on crystal contraction during heating or cooling due to neighboring crystals. The displacement of diffraction maxima and the broadening of diffraction peaks serve as indicators of micro stresses [2].

X-ray diffraction techniques offer remarkable sensitivity, enabling the measurement of lattice spacing, with precision up to one part in 10^5. This level of sensitivity allows for the detection of lattice strains as small as 10^{-5}. By leveraging the knowledge of the elastic modulus of a material, such as an aluminum alloy with a modulus of 60 GPa, it becomes feasible to calculate the corresponding stress. Even stress levels as low as 1 MPa, which represents a small fraction of the yield stress, can be accurately measured using X-ray diffraction. However, it is crucial to consider that X-ray diffraction measurements primarily capture information from the surface layers of the sample, and the measured values may not reflect the stress state across the entire material. Nevertheless, these calculations and measurements can be highly valuable in scenarios where assessing and controlling the quality of surface coatings holds significance.

5.3.2 Single crystal Laue diffraction

Accurate measurement of intensity of the diffracted X-rays is crucial in evaluating the effectiveness of the powder method in X-ray diffraction and identifying the lattice planes responsible for the formation of diffraction peaks. Alternatively, the spatial distribution of the diffracted intensity from a solid sample can be analyzed using a white radiation beam. The choice of wavelength depends on the orientation of the diffracting planes within the lattice structure, in accordance with Bragg's law. Interpreting the diffraction patterns from polycrystalline materials can be challenging due to the overlapping reflections from multiple sets. However, for single crystals, a distinct Laue pattern is produced, allowing for the determination of the crystal's orientation in relation to the incident beam.

The Laue camera, equipped with an areal detector array, can capture the diffraction pattern produced by a single crystal. Depending on the thickness of the sample, there are two possible camera configurations. For thin samples, the Laue pattern is recorded on a plane that is perpendicular to the incident beam. The Laue reflections from a set of crystal planes intersect at the same zone of symmetry, creating an elliptical pattern on the film plane. In a broader sense, the Laue diffraction patterns are documented as reflections. In contrast, there are no limitations on the thinness of the diffracting crystal in general. The incident beam passes through a cavity in the center of the recording plane, which is also perpendicular to the incident beam. The zones of symmetry associated with these reflections intersect the recognition plane as arcs of hyperbolas.

Identifying the symmetry axes associated with the reflection zones is crucial for interpreting the Laue pattern and determining the orientation of the single crystal with respect to the external coordinate system. Additionally, the Laue pattern offers valuable information about the quality of the single crystal. It can reveal additional sets of diffracted beams caused by twinned regions or sub-grains, which exhibit displacements compared to the diffracted beam originating from the single crystal [2].

5.3.3 Rotating single crystal methods

In certain cases, lattice models are necessary to validate the crystal model with a high level of certainty. This additional information complements the analysis of the crystallographic structure, which can be achieved by randomly orienting powder or polycrystalline samples. To conduct such an analysis, crystals are mounted at the center of the spectrometer or goniometer stage, and monochromatic radiation is utilized. The crystal axes can be accurately oriented relative to the spectrometer using the goniometer stage. Furthermore, the crystal structure can be continuously rotated around an axis perpendicular to the plane formed by the detection system and the incident beam. This rotation aligns the lattice points parallel to the spectrometer's plane, satisfying the reflection condition and resulting in a series of reflection lines.

Precise examination of the structure in the complementary space can be accomplished by employing adequately sized single crystals. It is advisable to work with crystals ranging from 0.1 mm to 1 mm in size to achieve the required resolution for a detailed structural analysis. To prevent overlapping patterns, it is crucial to restrict the rotation angle during data collection. Furthermore, it is necessary to record multiple spectra by rotating the crystal around significant symmetry axes. By diligently managing these parameters, a thorough and accurate analysis of the crystal structure can be obtained [2].

5.4 Diffraction analysis

In addition to the geometric aspects of diffraction, the angular distribution of the diffracted beams can be harnessed to analyze the crystallographic structure and determine relevant parameters, enabling the identification of crystalline phases within the crystal. However, valuable information can also be derived from the relative intensities of the diffracted beams. Examining these intensities entails considering various factors, such as the scattering caused by the individual atoms, the summation of the scattering contributions from individual grains and the atomic crystal, and the detection of the diffracted radiation by the diffractometer setup. By incorporating these factors, a more comprehensive comprehension of the crystal structure and its properties can be achieved.

5.4.1 Atomic scattering factors

In the following discussion, our focus will be on the interaction between the incident X-rays and the electrons, with the exclusion of the diffraction of the incident neutrons. The determination of the dispersed intensity can be accomplished by utilizing eq. (5.10), which is expressed as follows:

$$I = I_0 \frac{e^4}{r^2 m^2 c^4} \sin^2 \alpha \qquad (5.10)$$

The equation, which incorporates charge and mass of the electron (e and m, respectively), the distance between the accelerated electron and the incident beam (r), and the angle (α) between the scattering direction of the incident beam and the direction of the accelerated electron, appears to provide a description of the scattering of X-rays or electrons by an accelerated electron [2].

To accurately consider the impact of electromagnetic wave components on the electric field when dealing with an unpolarized X-ray beam, it becomes necessary to employ an averaging process. The electric field is perpendicular to the scattering plane xz and lies within the yz plane, resulting in the presence of two components, E_y and E_z. In the case of an unpolarized X-ray, these components exhibit an average equality. Additionally, the scattering direction forms angles $\alpha_y = \pi/2$ and $\alpha_z = (\pi/2 - 2\theta)$. with the electric field components. To account for equal contributions from each component to the total intensity, the factor $sin^2\alpha$ in eq. (5.10) needs to be replaced by $(sin^2\alpha_y + sin^2\alpha_z)/2$. Consequently, eq. (5.10) can be appropriately reformulated:

$$I = I_0 \frac{e^4}{r^2 m^2 c^4} \cdot \frac{(1 + \cos^2 2\theta)}{2} \qquad (5.11)$$

wherein the term $(1 + cos^2 2\theta)/2$ is referred to as the Lorentz polarization factor.

Every atom, characterized by its atomic number Z, contains Z electrons. Upon interaction with the incident beam, all electrons within the atom scatter coherently in a unified direction. The atomic scattering factor $f(\theta)$ quantifies the ratio of the amplitude of light dispersed by a single atom to the amplitude of the dispersed electron. As a result, for $\theta < 0$, the value of $f(\theta) < Z$ because the surrounding electrons of the atom scatter out of phase at larger scattering angles [2].

5.4.2 Scattering associated with the unit cell

The subsequent stage entails determining the amplitude, linked to the scattering originating from the crystal structure's unit cell. The pathway alteration (6) between the two beams, specifically the dispersed X-ray from an atom at the origin and the one dispersed by another atom within the crystal structure, corresponds to the phase difference (Q) between these dispersed beams. This phase difference is calculated as $\varphi = 2\pi\delta/\lambda$, where λ represents the wavelength of the X-ray.

Consider the location of the second atom that is well defined by the trajectory r. The coordinates of atom within the unit cell are represented by (x,y,z) and the direction of the position vector $[uvw]$ is determined by $u = x/a$. $v = y/b$ and $w = z/c$ The atom responsible for scattering the light resides on the plane denoted by the Miller indices (hkl), which is defined by the vector g, representing the complementary lattice vector.

As a result, the phase difference for the dispersed radiations can be determined using the eq. (5.12) provided below [2]:

$$\phi_{hkl} = 2\pi(hu + kv + lw) = 2\pi g \cdot r \qquad (5.12)$$

The amplitude A of the light dispersed by an atom is influenced by the atomic scattering factor $f_{z\theta}$. We can express the segment and the largeness of the dispersed wave using a vector A. To calculate the contribution of the atom at uvw to the largeness dispersed into diffracted ray hkl, we can utilize eq. (5.13) as follows:

$$Ae^{i\varphi} \propto f \exp[2\pi i(hu + kv + lw)] = f \exp(2\pi i g \cdot r) \qquad (5.13)$$

By disregarding the constant of proportionality in the previous formulation, it is possible to define a normalized scattering factor for the entire crystal's unit cell concerning reflections from the hkl plane. This can be expressed as shown in the following eq. (5.14):

$$F_{hkl} = \sum_{1}^{N} f_n \exp[2\pi i(hu_n + kv_n + lw_n)] = \sum_{1}^{N} f_n \exp(2\pi i g \cdot r) \qquad (5.14)$$

5.4.3 Interpretation of the intensities diffracted

Apart from the Lorentz polarization factor and the structure factors, various significant physical parameters contribute to determining the intensities of a deflection top, observed in a noted diffracted spectrum.

One of important factors contributing to the identification of the diffraction peak intensity is the multiplicity (P) associated with the reflecting planes. The multiplicity represents the numeral of planes fitting to a specific Miller indices group. For example, in cubic crystals, the multiplicity is 24 for planes within the unit triangle and 12 for planes along the edges of the unit triangle. Similarly, the multiplicity is 4 for planes in the stereogram and 3 for reflections corresponding to poles on the coordinate axes, collective by eight unit triangles.

Another physical factor that affects the intensity determination is the specimen dimension. In the dust technique, the accumulator only samples a portion of the cone with a finite cross section, diffracted at the Bragg angle. The Lorentz factor accounts for this fraction of collected radiations and is given by the following equation:

$$L = \frac{1 + cos^2 2a cos^2 2\theta}{sin^2 \theta cos\theta (1 + cos^2 2a)} \qquad (5.15)$$

where the diffraction angle, θ, is determined by the Bragg's law and the diffracting angle associated with the diffractometer is denoted by a.

Absorption effects in X-ray diffraction are influenced by the size and geometry of the sample. As the diffraction angle rises, the absorption becomes more significant. To correct for absorption, an absorption correction factor denoted as $A'(\theta)$ can be calculated based on the sample's dimension and density. In the case of thick specimens, the absorption factor is angle-independent and can be expressed as $A' = 1/2\mu$, where μ represents the linear absorption coefficient. When normalizing the calculated integrated intensities, the absorption factor is eliminated, ensuring precise and accurate measurements.

Temperature plays a crucial role in X-ray diffraction experiments. At elevated temperatures, the random vibrations of atoms disrupt the coherence of scattering. Coherence is a characteristic of closely spaced crystal planes with high diffraction angles. This effect becomes more pronounced for smaller lattice spacings and higher values of $\sin\theta/\lambda$. Moreover, coherence scattering is particularly significant for larger hkl values. The integrated intensity from an isotropic crystal is affected by the term: $e^{-2B\sin^2\theta/\lambda^2}$, where $B = 8\pi^2\bar{U}^2$ and $\bar{U}^2$ represents the average squared dislocation of each atom within the crystal.

Therefore, with due consideration to the aforementioned discussed factors, a general association for the diffracted intensity can be obtained through the following formulation:

$$I = k|F^2| \frac{1 + \cos^2 2a \cos^2 2\theta}{\sin^2\theta\cos\theta(1 + \cos^2 2a)} PA(\theta)\exp\left(-\frac{2B\sin^2\theta}{\lambda}\right) \qquad (5.16)$$

The scaling factor, represented by k, incorporates the term 10. By utilizing eq. (5.16), the calculated integrated intensities for a specific crystal structure can be simulated. Subsequently, the resulting integrated intensities are obtained for each reflection corresponding to the hkl planes. To ensure consistency, these derived intensities are normalized by dividing them by the maximum derived intensity, as depicted in eq. (5.17):

$$I_{hkl}^n = \frac{I_{hkl}}{I_{hkl}^{max}} \times 100 \qquad (5.17)$$

5.4.4 Errors and assumptions made

At this juncture of the chapter, it is essential to address the errors associated with the intensity measurements. These errors can be categorized into two distinct types: those related to the dimension of the top intensity as well as those associated with the identification of the top position. To ensure accurate resolution of small changes in lattice parameters, proper instrument calibration is crucial. X-ray diffraction techniques can be particularly advantageous in situations where only a small volume of the material is sampled, making them well suited for the study of slim layer, and solid- like instruments. Various systems, including microelectronic components, where parts, and optoelectronic devices, greatly benefit from such meticulous investigations. X-ray diffraction methods enable the monitoring of surface changes and chemical composition within the lattice. To meet the measurement requirements under high-temperature conditions and controlled environments, advanced X-ray diffraction instruments are readily available in the market.

The discussion offers a concise overview of the errors associated with the measurement of lattice parameters, with particular emphasis on the widely used and automated Bragg-Brentano diffractometer. This instrument configuration involves positioning the focal circle in a way that it touches the sample surface as well as the X-ray source. As specimen angle of rotation (θ) increases, the radius of the focal circle diminishes. To minimize the broadening of the signal beam, it is crucial to align the detector system with the varying focal circle. Adhering to these operational conditions helps to manage three key errors in the diffraction peak. However, the failure to comply with these conditions can impede the accuracy of the measured lattice parameters and lattice spacings.

The initial form of the error is peak broadening, which arises when there is inadequate alignment of the diffractometer or when the specimen is incorrectly positioned on the goniometer. Enhancing the alignment or utilizing narrower receiving slits can help minimize peak broadening by restricting the dispersion of the diffracted signal. Nonetheless, despite improved alignment and the use of smaller slits, a certain degree of residual broadening may persist, resulting in a fault in the measured diffraction angle.

The remaining two errors are associated with the sample's sizes. Firstly, the deflection indicates rises from an area below the surface edge of the specimen, rather than directly from the surface edge itself. This discrepancy introduces a fault in the perceived deflection angle. Secondly, if the sample is positioned either up or down the pivotal point along the goniometer axis, a similar error occurs, referred to as the defocus error. The magnitude of this error can be determined using the following equation:

$$\frac{\sin \Delta 2\theta}{h/\sin \theta} = \frac{\sin(180 - 2\theta)}{R} \tag{5.18}$$

The above can be simplified as follows:

$$\Delta 2\theta \cong \frac{-2h \cos \theta}{R} \tag{5.19}$$

Finally, the relationship can be depicted in eq. (5.20):

$$\frac{\Delta d}{d} = \frac{-\Delta 2\theta}{\tan \theta} = \frac{\cos \theta}{\sin \theta}\frac{2h \cos \theta}{R} = k\frac{\cos^2 \theta}{\sin \theta} = \frac{\Delta a}{a} \tag{5.20}$$

To minimize faults in the measurement of the lattice parameters, several approaches can be employed. One approach involves extrapolating the calculated values from the observed diffracted peaks to a value corresponding to $\theta = 90^{\circ}$. This can be achieved by plotting the lattice parameters as a function of $cos^2\theta/sin^2\theta$ and extrapolating the plot to 0. Another approach is to utilize a standard powder with known lattice parameters and apply it to the surface edge of the sample. The peak positions derived from the standard powder can be utilized to correct systematic errors. These methods contribute to the reduction of errors and enhance the accuracy of the measured lattice parameters.

When measuring the peak intensities in X-ray diffraction, it is essential to consider additional factors. One crucial factor is the geometry of the specimen within the spectrometer, as the finite volume of the specimen can result in a geometrical peak broadening, which must be accounted for. Another factor to consider is the reply of the detection system to the incident photons on the specimen. It is of utmost importance to ensure that the detection system remains consistent and reliable throughout prolonged periods of operation. By taking these factors into consideration, it becomes possible to obtain accurate and reliable peak intensities in X-ray diffraction measurements.

Comparative intensity measurements of different diffraction peaks play a crucial role in identifying unknown phases and refining the crystal structure models. However, it is important to consider the influence of preferred orientations on intensity measurements. To ensure accurate results, it is recommended to measure the integrated peak intensity rather than solely relying on the peak height in the diffraction spectrum. By calculating the area under the peak and subtracting the background noise, integrated peak intensities can be obtained. This approach yields more reliable and meaningful data for analysis and interpretation purposes.

When dealing with a thin sample, it is important to account for both the preferred orientation and the temperature variation when calculating the corrected integrated intensity. Furthermore, it is essential to consider the path length while multiplying the integrated intensity to ensure accurate representation of the scattering contribution from the sample. By applying this correction, the measured intensity can accurately reflect the sample's characteristics and provide more reliable data for analysis and interpretation purposes:

$$[1 - \exp(-2\mu t \cos ec\theta)] \tag{5.21}$$

For a thick specimen, the factor mentioned above is equal to I. However, when the absorption coefficient (μ) is known, it is possible to determine the thickness (t) of a thin sample by a comparison of the intensities with those obtained from a bulk material. By using the refinement procedure and obtaining the relative integrated intensities, the product of μ and t can be accurately determined. Another approach is to compare the intensities of the two reflections recorded at different Bragg angles, θ_0 and θ_1. The following expression can be used for this purpose:

$$Y = \frac{I(t)}{I(\infty)} = \frac{1 - \exp(-2\mu t \cos ec\theta_1)}{1 - \exp(-2\mu t \cos ec\theta_2)} \tag{5.22}$$

The aforementioned method for determining the sample thickness is practical only when the product μt ranges 0.01–0.5.

The relative integrated peak intensities are affected by the preferred orientation of the sample, which provides valuable information about the microstructural features that play a significant role in its mechanical properties. Although a comprehensive explanation of the precise techniques for determining the preferred orientation is beyond the scope of this discussion, there is a method called the Harris method that enables a qualitative assessment of the texture. This method involves evaluating the size portion of the phase with a specific crystal alignment within a minor solid angle $d\Omega$, as defined by eq. (5.23):

$$P(\alpha, \beta, \gamma)d\Omega/4\pi \tag{5.23}$$

The function $P(\alpha, \beta, \gamma)$. depends solely on the values of the angles α, β, γ. Meanwhile, each crystal possesses some alignment and the following condition will always be true:

$$\frac{1}{4\pi} \iint P(\alpha, \beta, \gamma)d\Omega = 1 \tag{5.24}$$

For the case of materials that are oriented randomly and is independent of the angles, the following relation will hold true:

$$\frac{P}{4\pi} \iint d\Omega = 1 \tag{5.25}$$

In the case of a polycrystal with random orientation, the probability factor P is equal to 1. When a crystallographic direction has a higher likelihood, the value of P can exceed 1, indicating a higher probability for that direction. Conversely, if the value of P is less than 1, it suggests that a particular crystallographic direction is less likely to be present. The Bragg-Brentano diffractometer is commonly used in experimental studies to determine the value of P:

$$P(\alpha, \beta, \gamma) = \frac{I(hkl)}{\sum I(hkl)} = \frac{\sum I'(hkl)}{I'(hkl)} \tag{5.26}$$

In the equation provided, $I(hkl)$ represents the measured intensity obtained from a plane characterized by the Miller indices (hkl), while $I'(hkl)$ represents the calculation of the integrated intensity originating from a similar plane but with random orientation.

In quantitative analysis, the measured intensity plays an important part in evaluating both the degree of texture on the chosen diffracting plane and determining the phase composition within the specimen. This analytical technique is essential for understanding how various dispensation limitations impact the phase contents. To quantify the proportion of each phase, such as α in a combination of α and β, the combined power is defined by employing eq. (5.27):

$$I_{\alpha} = \frac{K_1 c_{\alpha}}{\mu_m} \tag{5.27}$$

In eq. (5.29), the intensities of phase a within the multiphase material are symbolized as c_{α}, while the linear coefficient of absorption for the mixture of phases is denoted as μ_m. Additionally, Kl represents an arbitrary constant. The coefficient of linear absorption is affected by the relative quantities of the constituent phases within the multiphase material. By utilizing eq. (5.29), a correlation is established to calculate this coefficient:

$$\frac{\mu_m}{\rho_m} = \omega_{\alpha}\left(\frac{\mu_{\alpha}}{\rho_{\alpha}}\right) + \omega_{\beta}\left(\frac{\mu_{\beta}}{\rho_{\beta}}\right) \tag{5.28}$$

where ω is the weight-fraction that is associated with each of the constituting phases in the multiphase material, and the density of the multiphase material is denoted by ρ. Rearrangement of eq. (5.28) yields eq. (5.29) as follows:

$$I_{\alpha} = \frac{K_1 c_{\alpha}}{c_{\alpha}\left(\mu_{\alpha} - \mu_{\beta}\right) + \mu_{\beta}} \tag{5.29}$$

Equation (5.30) provides a relationship between the phase a in the phase mixture to the present in the pure sample:

$$I_{\alpha} = \frac{K_1 c_{\alpha}}{c_{\alpha}\left(\mu_{\alpha} - \mu_{\beta}\right) + \mu_{\beta}} \tag{5.30}$$

If the mass absorption coefficients are known for each of the constituent phases in the multiphase material, eq. (5.30) provides a satisfactory approach. However, when the mass absorption coefficients are unknown, the construction of a calibration curve necessitates the preparation of a set of standard specimens. Regardless of the scenario, it is advisable to perform experimental calibration to ensure accurate results.

While there are various diffraction techniques available, such as residual stress analysis, particle size determination, and thin-film X-ray techniques, these methods extend beyond the scope of the current discussion.

5.5 Illustrative example

This section of the chapter focuses on the X-ray diffraction analysis of doped Zr02 powders with nanocrystalline structure. The powders were synthesized using the gel combustion method, utilizing critic acid as a fuel source. Detailed information on the synthesis method can be found in the works by Juarez et al. and Lamas et al. [11–13].

Figure 5.1 presents the X-ray diffraction pattern of the Y_2O_3-doped ZrO_2 powders. It highlights the retention of the tetragonal phase. Notably, the diffraction pattern does not exhibit peaks corresponding to the stable phase. These findings indicate that only in non-crystalline samples is the presence of the tetragonal phase observed while the monoclinic phase is absent [11–13].

To determine the average size of the crystal, a symmetric Pearson VII function was employed to fit the profile peak $(1\,1\,1)_t$. This fitting procedure facilitated the calculation of the integral breadth. Scherrer's equation was then utilized to determine the average size of the crystal. It is important to note that the effects of micro strains and instrumental broadening were not considered during these calculations. The calculated value of the average size of the crystal was determined to be 92(2) Å.

The obtained estimate in the given scenario provides a quantitative value with a certain level of accuracy. This accuracy can be attributed to the fact that strain-induced broadening is not significant at low angles. The broadening of the Bragg peaks is primarily influenced by the tangent of the diffraction angle i.e., $\tan\theta$. Moreover, the intrinsic profile, which represents the true shape of the peaks, is minimally affected by instrumental broadening. This observation is evident in Fig. 5.1, where the diffraction peaks exhibit relatively narrow profiles. The instrumental broadening, typically within the range of a few hundredths and with a maximum value of approximately 0.1° degrees, has a limited impact on the overall profile.

For the analysis, the Reitveld method, in conjunction with the conventional pseudo-Voigt function, was employed, and the results obtained are presented in Tab. 5.1. The analysis yielded an estimated value of $<D>_v = 120(5)$ Å, representing the average crystal-

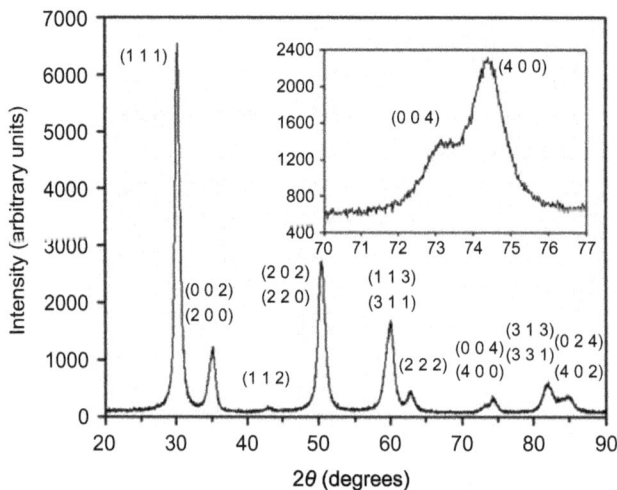

Fig. 5.1: Diffractogram of Y_2O_3-doped ZrO_2 powder (reproduced with permission from [12]).

lographic dimension. Table 5.1 provides the determined parameters pertaining to the crystallographic characteristics that were obtained through the fitting procedure.

Tab. 5.1: Estimated values of average crystallite size (reproduced with permission from [12]).

β_L^S (degree)	β_G^S (degree)	β_L^D (degree)	β_G^D (degree)	β^S (degree)	β^D (degree)	$<D>_v$ Å	$\tilde{e}(\times 10^3)$
0.33	0.50	1.12	–	0.73	1.12	120(5)	4.9(2)

The calculated parameters obtained from the conventional pseudo-Voigt function were used to determine the integral breadths of the diffraction peaks. Figure 5.2 illustrates the Williamson-Hall plot, cosθ vs Sino, constructed using the obtained parameter values. During the calculation of the parameter values for ß, no instrumental corrections were applied, but the effects of Ka2 radiation were considered. From the plot in Fig. 5.2, it can be observed that the data points approximately align along a slightly curved line, indicating the dominance of the Lorentzian component in the obtained profile. The steep slope value suggests a significant influence of strains. Additional estimates were derived based on the line intercepts and slope of the line, and the determined values are presented in Tab. 5.2.

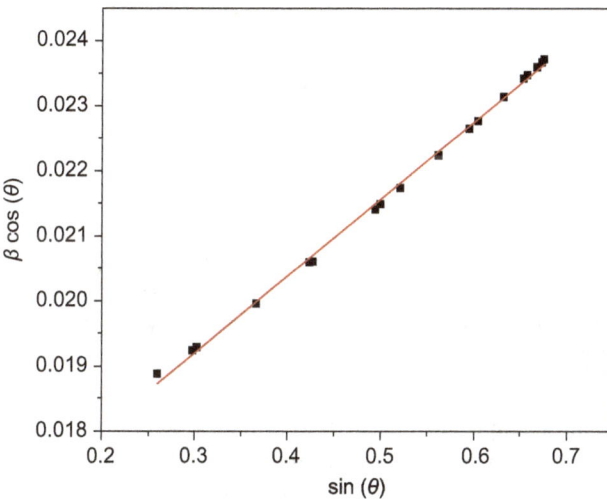

Fig. 5.2: Williamson-Hall plot (reproduced with permission from [12]).

Tab. 5.2: Estimated values of average crystallite size obtained from Williamson-Hall plot (reproduced with permission from [12]).

$<D>_v$ Å	$\tilde{e}$ $(\times 10^3)$
98(1)	2.95(5)

5.6 Conclusion

When short-wavelength radiations, such as X-rays, electrons, or neutrons, interact with the ordered arrangement of atoms in a crystal structure, they undergo elastic scattering. The scattering occurs at specific angles determined by the incident angle and the crystal structure. The intensity and angle of scattering are influenced by the properties of the incident radiation, the crystal structure, and the wavelength of the radiation. By capturing and measuring the dispersed intensity in space, a diffraction pattern is generated. This can be accomplished using photographic emulsion or suitable charge-gathering devices. The diffraction pattern represents the intensity of the diffracted radiation collected at different scattering angles.

Laue's work demonstrated that the angles required for scattering from a single crystal can be determined, based on the lattice parameters of its unit cell. Building upon this, Bragg further simplified the relationship, giving rise to the Bragg's Law. According to the Bragg's Law, the wavelength of the incident radiation is twice the space between the crystal planes multiplied by the sine of the scattering angle. Thus, the incident wavelength, plane spacing, and scattering angle are interconnected. The scattering angles that satisfy the Bragg's Law are known as the Bragg diffraction angles and agree to integer standards of the Miller indices. However, not all mixtures of the Miller indices result in observable diffraction peaks. In certain cases, specific reflections are considered "forbidden" and can be disregarded. This typically occurs when the Bravais lattice contains body- centered or face-centered lattice-points.

The intensities of the diffracted top are gritty by both the atomic numeral of the atoms and their positions within the crystal lattice. Individually, a lattice point is associated with specific atoms that belong to the space group. When these atoms scatter in-phase with each other at their respective lattice points, it can result in enhanced diffraction peaks. Conversely, if the scattering is not in sync, destructive interfering may reduce the observed intensities of the peaks. Thus, the relative positions and the scattering phases of the atoms play a crucial role in determining the intensity of the diffracted radiation.

A complementary space serves as a valuable framework for representing angular positions of permissible diffraction peaks. Within a complementary space, the complementary lattice encompasses lattice points that correspond to the positions of diffracting planes. These lattice points are associated with the complementary interplanar space of the deflecting planes and the course of the vector perpendicular to the plane. The Bragg calculation establishes a sphere in the complementary space and the complementary lattice point that somewhat satisfies this equation will generate a diffracted beam upon reflection from the sphere. By utilizing the complementary space, we can conveniently comprehend and analyze the phenomenon of diffraction.

As the incident beam's wavelength decreases, the reflecting sphere's radius in the complementary space expands. Consequently, a greater number of complementary lattice points, each associated with distinct diffraction conditions, have the opportunity to intersect with the surface of the reflecting sphere. Hence, smaller grain sizes or crystal rotations heighten the chances of complementary lattice points intersecting the reflecting sphere's surface, leading to enhanced diffraction. This phenomenon elucidates why finer crystalline structures or crystal rotations contribute to intensified diffraction peak intensities within a diffraction pattern.

An X-ray diffractometer comprises several crucial components necessary for its operation: an X- ray source, a goniometer stage, a detection system, and a data recording system. The X-ray source is responsible for producing X-rays that are then directed toward the sample under investigation. The goniometer stage allows the precise manipulation of the sample's position and orientation, facilitating the selection of specific diffraction angles for analysis. The detection system plays a vital role by capturing the diffracted X-rays and converting them into electrical signals for further processing. This system can take the form of a single detector, photographic emulsion, or an array of position-sensitive detectors, depending on the experimental setup. In certain cases, a Laue camera is employed to record the diffraction patterns from single crystals, using white X-ray radiation. On the other hand, for polycrystalline samples, X-ray diffraction studies utilize monochromatic radiation to obtain powder diffraction spectra, which find wide application in structural analysis and phase identification across various fields.

Part B: **Spectroscopic instruments**

Chapter 6
Fourier transform infrared (FT-IR) spectrometer

6.1 Introduction

Since the discovery of infrared radiation in 1800 by Sir William Herschel, various techniques and instrumental advancements have been made in measuring the infrared spectra. A prism and a mercury thermometer were first used by Herschel to make observations on heat-based radiation beyond the solar spectrum. Melloni then discovered the transparency of NaCl to infrared radiation and later invented the first midinfrared spectrometer in 1833. It was only with the invention of the interferometer (for measurement of the speed of light in diverse directions) by Michelson in around 1880s that a fully operable spectroscopy instrument came to be used widely. Modern infrared spectroscopy with the introduction of Fourier transform by Biorad Company of Cambridge, Massachusetts, in 1969 has again enabled the use of the instrument as a characterizing tool worldwide. Developments have also been made in the components of the infrared spectroscopy, such as the excitation source (He-Ne laser source), advanced infrared detectors, analog to digital converters, and minicomputers.

A detailed description of the instrumental details of an FT-IR spectroscopy [14, 15] has been provided in the chapter. The fundamental knowledge of the instrument that is required by any FT-IR spectrometrist regarding the sample preparation, data handling, and interpretation has been provided.

6.2 Instrumental details and working

In order to study the infrared spectra of vibrational molecules, FT-IR spectrometer can be majorly divided into its components as: the radiation source, interferometer, detectors, and optical instruments (mirrors, beam splitters, etc.). These components have been discussed in detail.

6.2.1 Radiation source

Both dispersive and Fourier-transformed instruments are suitable for using the same type of infrared sources. Hence, the use of any infrared source is not influenced by their instrumental setup. Commonly used sources of radiation includes Globar source, Nernst glower, nichrome coils (Fig. 6.1) and mercury arc lamps, Mercury arc lamps (Fig. 6.2) have been used most commonly in the far infrared region of the spectrum. Nichrome coils have low emissivity and operate at low temperature.

https://doi.org/10.1515/9783110997590-006

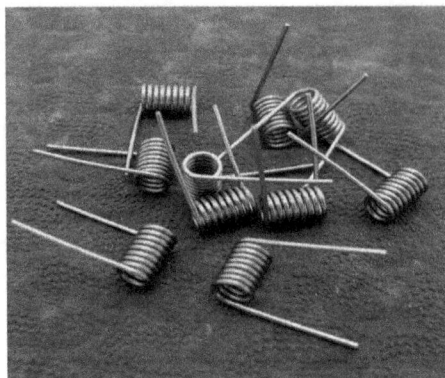

Fig. 6.1: Nichrome coils.
Source: Nichrome coils (2020) (reproduced with permission from [16]).

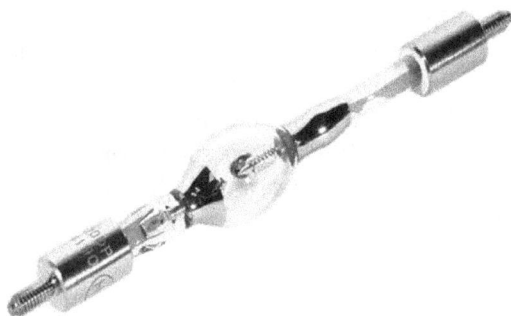

Fig. 6.2: Mercury arc lamp (right).
Source: Mercury arc lamp (2019), lightning, and ceiling fans (reproduced with permission from [17]).

6.2.1.1 Globar sources

Globar sources (Fig. 6.3) are made out of SiC with metallic leads on both ends. When voltage is applied across, the source is heated to a high-temperature-emitting radiation at a temperature above 1000 °C. Due to their high emissivity even at a frequency of a wave number of 80 cm^{-1}, they are excellent sources of far infrared region of spectrum. This high temperature emission directs the need to cool the metallic electrodes using water.

6.2.1.2 Nernst glowers

Nernst glowers (Fig. 6.4) are made from ceramic sources, which remain nonconducting at room temperature and behave like a conductor after intense heating. The ceramic element of the heating source is a mixture of yttrium and zirconium oxides and

Fig. 6.3: Globar source.
Source: Globar source of Paris (2016), https://eureka.physics.utoronto (reproduced with permission from [18]).

is air-cooled rather than by water as in case of Globar sources. Due to such technique of cooling, Nernst glowers are easy to maintain and portable. The emission spectrum of the source at 1800 K approaches that of the black body. The disadvantages are its mechanical instability and low life.

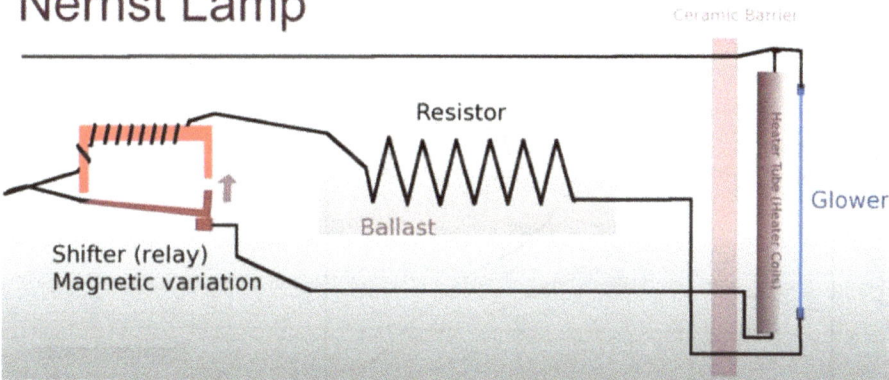

Fig. 6.4: Nernst glower.
Source: Nernst lamp graphics (2010), Edison Tech Center (reproduced with permission from [19]).

6.2.2 Infrared detectors

Infrared detector is one of the most important elements of infrared spectroscopy, the performance of which depends not only on its ability to capture photons and produce electrical output but also on aligning the various optical components like mirrors, windows, apertures, Dewar flasks, etc. Infrared detectors can be classified into two main types, which are thermal detectors and photon detectors. Thermal detectors are sensitive to the temperature fluctuations in a sample due to the incoming beam of infrared light causing the material to heat up at the incident region. Photon detectors are sensitive to the change in intensity or the quantity of photon particles from the irradiated sample. These are described in detail below:

6.2.2.1 Thermal detectors
Thermal detectors rely on the changes in electrical or physical characteristics (such as electrical resistance, thermovoltaic effect, electrical polarity, and thermopneumatic effect) of various elements with fluctuations in temperature of the sample being irradiated with infrared beam. Four different processes are generally employed for thermal detection of the samples irradiated with infrared radiation. These are bolometric effect, thermovoltaic effect, pyroelectric effect, and thermopneumatic effect. *Bolometric effect* generally involves the change in electrical resistance with minute rise in temperature caused as a result of the samples being irradiated with infrared radiation and absorption of the wavelength in the infrared region. *Thermovoltaic effect* is based on the generation of voltage across the junctions of two dissimilar metals due

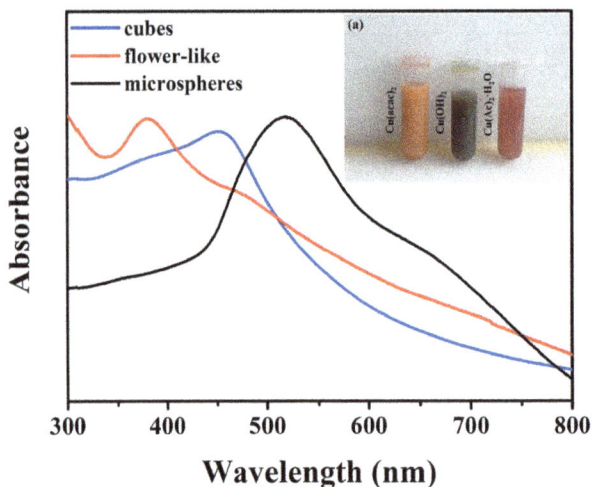

Fig. 6.5: Golay detectors.
Sources: Schematic of Golay cells (2011) (reproduced with permission from [20]).

to rise in temperature between the junctions. *Golay detectors* (Fig. 6.5) are generally based on the thermopneumatic effect, involving absorption of the infrared radiation by a sample element placed inside a gas filled chamber with a deformable and flexible mirror on one side and optical window on the other side. The distortion in the flexible membrane is sensed by an independent optical system, which enables radiations at the far infrared region of the spectrum to be detected conveniently. Pyroelectric effect is another phenomenon, which is based on the generation of electric field due to polarization of the molecules on the surface of the sample material irradiated by infrared source.

6.2.2.2 Photon detectors or quantum detectors
Photon detector (Fig. 6.6) or quantum detectors are generally based on change in number of charge carriers with a number of infrared photons striking the sample. The quantum-level phenomena that enable a photon detector to work can be classified into four types, which are photoconductive effects, photovoltaic effects, photoemissive effects, and photoelectromagnetic effect. *Photoconductive effect* works on the principle that when an infrared photon is allowed to strike off electrons from the holes present in a silicon-based semiconductor, charge carriers (electron hole pairs) are generated. With more photons striking the surface, more number of charge carriers is produced causing an electric current to flow and making it conductive. *Photovoltaic* effect is a similar effect where photons striking the *p-n* junction cause a voltage difference across the junction. In *photoemissive effect*, photon energy is transferred to free electrons on the surface of the sample kept in an evacuated chamber with an anode placed nearby. The emitted electron is picked up by the anode to generate a current that is detected. In *photoelectromagnetic effect*, a magnetic field is applied to separate charges that induce a voltage across, which is proportional to the amount of photons received.

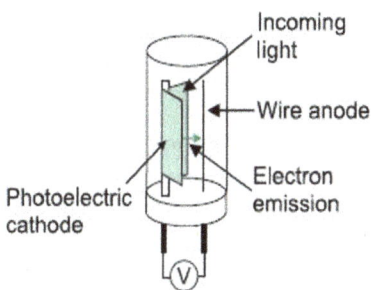

Fig. 6.6: Photon detector.
Source: Photon detectors (2006), SPIE Press, Bellingham, WA (reproduced with permission from [21]).

6.2.2.3 Mercury Cadmium Telluride
Mercury Cadmium Telluride (MCT) detectors (Fig. 6.7) are the most widely used detectors, which are basically photon detectors and need to be operated at a low temperature of 77 K using liquid nitrogen as the cooling agent. The spectral response of the MCT detectors generally depends on the composition ratio of the HgTe and CdTe, which influences

Fig. 6.7: MCT detector.
Source: MCT cryogenic receiver (2020), Sciencetech (reproduced with permission from [22]).

the wavelength at which the spectrum result is in compliance with the energy gap between different energy levels in the molecules. Therefore, for different compositions, spectral response is the most suitable for a specific wavelength of the incoming radiation.

6.2.3 Michelson interferometer

Michelson interferometer (Fig. 6.8) is the first instrument used to analyze the intensity of the output radiation in relation to the optical path difference, the principle of which is employed in all other modern interferometers. The produced interferogram is then interpreted with the intensity of the radiation as a function of the frequency of the radiating source.

It basically consists of a beam splitter, two perfectly flat mirrors, a detector, and a source. One of the flat mirrors is placed along a path (right angle to the incident radiation) from the beam splitter which is a fixed mirror. The other flat mirror is movable and placed at an angle of 180° to the incident radiation along the beam splitter. The radiation from the source first strikes the beam splitter at an angle of 45°, which is basically made out of a thin layer of germanium on a flat substrate of KBr (infrared transparent material). The beam then splits into two equal intensities of light and gets reflected back from the fixed and the movable mirror to form an interference pattern on the beam splitter. This interference pattern is perceived by the detector in the form of an interferogram. With change in the optical path length of the two beams reflected from the fixed and the movable flat mirrors, interference in intensities can be observed. For equal distance travelled by the reflected beams, interference is basically a constructive one, while with change in path travel distance of the two beams by about the multiples of 180° out of phase of the source wavelength, a destructive interference is observed. The final intensity on the detector as a function of optical

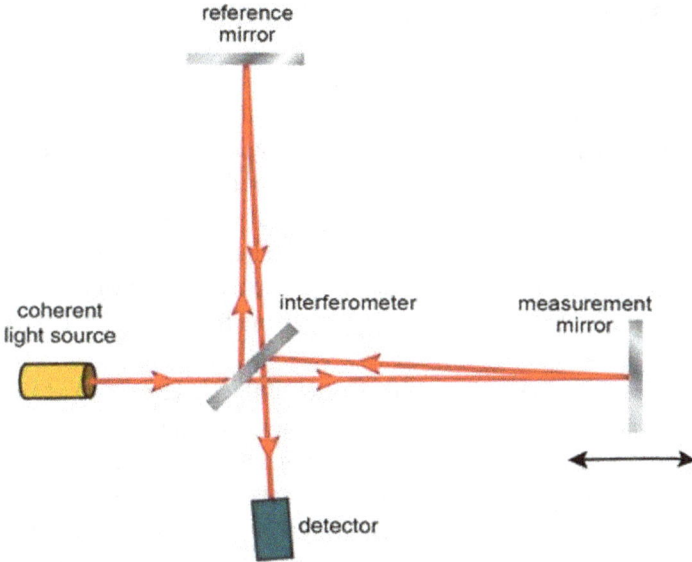

Fig. 6.8: Michelson interferometer.
Source: Michelson interferometer (retrieved on 25 February 2020) (reproduced with permission from [23]).

path difference can be given by a cosine relation with the intensity (as a function of the frequency of the source radiation) and the frequency itself as follows:

$$I(\delta) = I(v)\cos(2\pi v\delta) \tag{6.1}$$

where v' is the wave number of the source radiation and δ is the optical path difference. Fourier transform is applied to the above equation, which gives the intensity peaks as the function of the frequency of the source radiation.

   A few of the advantages of interferometers include the simultaneous detection of the whole spectrum leading to more multichannel advantages and achieving the desired signal to noise ratio speedily, throughput advantages over the infrared frequency range from a range of 10–250, and accurately and precisely indexing retardation using monochromatic source enabling determination of retardation sampling procedure.

6.3 Sample preparation

FT-IR samples can be prepared by two of the most popular sampling techniques: transmission sampling and reflectance sampling. Samples prepared using the transmission method are basically in the form of powders, liquids, or gas whereas it is in the form of bulk solids in reflectance sampling. In this section, transmission sampling has been discussed in detail and a general idea of the advantages and disadvantages of the reflectance sampling has been provided.

6.3.1 Transmission sampling

This method is extensively used in the preparation of almost all samples in FT-IR spectroscopy except for a few which do not yield usable transmission spectrum. The infrared radiation is allowed to pass through a thin layer of the sample having a thickness of about 1–20 microns for the final absorption spectrum lines on the interferogram. The advantages of transmission sampling are low tooling cost as compared to the reflectance sampling, good signal-to-noise ratio and resolution of the transmission spectrum, and trusted history and tradition in the results obtained. The major disadvantages of transmission sampling include the sample opacity problem and the destructive nature of the experiment.

Sample opacity is mainly dependent on the product of the concentration of the samples and their thickness (or the path length) through which the infrared is allowed to pass. With lower concentration of the sample and path length the source radiation will just pass through without any spectrum formation, whereas with larger value of the product of the path length and the concentration of the samples, infrared spectra exhibits a distorted one with more absorption by the sample taking place. The solution for this problem is to select the combination of concentration and path length in such a way that the absorption peaks are lesser than 2 absorbance units and greater than 10% T.

The destructive nature of the FT-IR sampling is another drawback of the transmission sampling technique, which basically implies that bulk solids need to be broken down into powdery form to be presented as an FT-IR sample. The operations involving grinding, compression, and dissolving the sample are critical and cannot be applied to samples that are valuable and need to be intact.

6.3.1.1 Transmission sampling of solids and powders

Transmission sampling of solids and powders can be done in two different ways: preparing potassium bromide (KBr) pellets and by Mull technique.

KBr pellets are generally used in FT-IR spectrometer due to their property of inertness and transparency to infrared radiations (above 400 cm^{-1}), keeping the sample spectral peaks intact on the interferogram. It is used to obtain the infrared spectrum of powder sample or any other solids that can be ground into powder. KBr and the sample powders must be ground to particles having sizes lesser than 2 microns to avoid infrared scattering effect of larger particles. Both the powder particles are ground in separate mortars so that reaction can be prevented under heat and pressure of the grinding. KBr is checked for a paste-like material and a dull appearance, which indicates reduction in scattered light intensities and reaching appropriate particle size. The mixture of both the ground materials is then thoroughly stirred with a spatula for about 1 min to maintain uniformity of the sample particles in the KBr paste. Pressure of about 5000–10,000 PSI is applied on the mixture of KBr and the sample powder in a "bolt press," which forms into a glass pellet embedded with sample particles. When exposed to infrared beam, these pellets with particles in them generate a spectrum on the interferogram.

The drawbacks of the KBr pellet is its hygroscopic nature, which results in absorption of water from the atmosphere and formation of a thick layer of water over the pellet. This drastically reduces the quality of the infrared spectra and blocks the absorption lines of the sample obtained on the interferogram. To prevent this, the pellets are generally heated in an oven or a desiccator at a temperature of above 100 °C to get rid of the water from the KBr sample. Another drawback of KBr is its water solubility, so any sample containing water cannot be analyzed using it.

Mull Technique uses powdery solids, which are produced by grinding the solid sample to reduce the effect of particle size on the infrared spectrum obtained. The solid particles are then mixed thoroughly with a drop (or two) of mineral oil or "mull" using a spatula until it transforms into a homogeneous paste or slurry. A little amount of this slurry is placed over an infrared transparent window (or KBr plate) and another window is placed over the slurry in a sort of "sandwich." It is generally preferred to press gently both the windows together to spread out the slurry and enhance the reproducibility of the windows by mechanical means. The capillaries present in the slurry sample bind both the windows together by capillary action. For the background spectrum, first the KBr windows are placed in the infrared beam without the slurry sample being loaded. Then, with slurry being pressed between the same windows the sample spectra are analyzed.

6.3.1.2 Transmission sampling of liquids

As the liquid samples are easier to spread onto the windows, transmission sampling of liquids are faster than in case of solids. Two methods are generally employed in the sampling, which are the capillary thin films method and sealed cells method.

6.3.1.2.1 Capillary thin films

In capillary thin films, a drop of the liquid sample is generally put over the infrared transparent window and another window placed over the first window. The windows are then rotated in opposite direction in order to uniformly spread the liquid whose capillary action holds both the windows together. The sample is then placed in the sample holder and an infrared beam is allowed to pass through it. As in the case of mull technique, background spectrum is first run with only the windows being placed in the infrared beam, and the same windows must be used to load samples for the second scan. Quality spectrum is obtained using the capillary thin film sampling technique with low noise, well-resolved peaks, flat baseline, and negligible offset. For the purpose of cleaning the thin film sample from the windows, the windows are generally rinsed with solvents such as ethanol, tetrahydrofuran, and methylene chloride that can dissolve the thin film sample but not the window. Semisolid samples, which are viscous, or soft solid can also be used in the same manner as discussed above for transmission spectrum analysis.

One of the major drawbacks of capillary thin film sampling is the inability to use water-soluble samples or aqueous solutions, which dissolve the infrared transparent KBr or NaCl windows. This problem can be solved by using water-insoluble windows such as ZnSe and AgCl but these are expensive. Anther limitation of the capillary thin film sampling is the non usability of volatile sample as it is not sealed between the windows and can easily evaporate during detection of the spectrum. Also the quantitation of the capillary thin film is very difficult as there is no provision for testing with different path lengths and lack of control over it, though qualitative analysis of organic samples can be performed very easily.

6.3.1.2.2 Sealed liquid cells

Sealed liquid cells overcome the drawback of using volatile (liquid tat evaporates) sample in the capillary thin film method. Here the sample is sealed inside a pair of windows separated by a gasket to form a cell. The windows are made out of KBr or any other infrared transparent materials and one of the two has two tiny holes on it. The sealed liquid cells have a thickness in the range of 1–500 microns and the gaskets are made of Teflon or lead/mercury amalgam. Liquid sample is inserted into the cell using a Luer-lock tip and fitting the combination by a syringe. The tip part is fitted onto the syringe while the fitting part is fitted to the cell holder. Two syringes are used, one to insert the liquid sample and the other to draw out the air inside the cell and in turn pulling the liquid into the cell. Once completely filled, the cell is checked in the light for any air bubbles left and in case there is air bubble the cell is refilled. The plugs inserted into the Luer-lock fitting after the filling of the liquid sample prevent evaporation of the liquid from the sample (through the holes) and air contact, and are made of Teflon.

6.3.1.3 Transmission sampling of gases

Gases and vapor can be appropriately analyzed using the transmission method. Due to higher dilution as compared to liquids and solids, the path length of the infrared generally opted for gas or vapor samples is more in the unit range of meters to kilometers for the transmission method. A commonly known "10-cm cell" is generally used to contain the gas at higher concentration of above 1% for transmission sampling.

In the "10-cm cell," as the name suggests, a 10-cm path length is used, which is basically made out of metal or glass tube on the sides and infrared transparent windows on both sides of the vessel. The windows are sealed to the cell body by gaskets and retaining rings and a pair of glass stoppages for valves protrudes out from the cell body on its upper part. The gas filled 10-cm cell is held in the infrared beam using cell holders that are mounted onto the sample compartment slide.

In order to load the gas into the 10-cm cell a special arrangement consisting of a vacuum manifold is attached to the stoppage valves, pressure gauge, vacuum pump, and the gas cylinder attached via plumbing. The gas released from the gas cylinder passes through the manifold into the gas cell. Before releasing the gas from the cylinder, the vacuum manifold is evacuated using the vacuum pump while the valve to the gas cell is still closed. After completely evacuating the manifold, the valve is opened to pull away the remaining air in the cell, creating vacuum in the cell. The vacuum pump is closed and gas is released from the gas cylinder via a valve into the manifold and then into the gas cell. The gas cylinder valve and the manifold valve are closed simultaneously and vacuum pump is again switched on to drive away the remaining gas in the manifold. The gas-filled cell is now ready to be mounted onto the cell holder and infrared passed for spectrum analysis. As in all other cases, the evacuated cell is first tested to obtain the background spectrum followed by the gas-filled cell.

6.3.2 Reflectance sampling

Reflectance sampling employs the principle of reflection of light where infrared radiation is focused onto the surface of the sample and the reflected beam is used for qualitative analysis of the spectrum obtained. As already known, reflectance phenomenon can be obtained in three different methods, that is, specular reflectance, diffuse reflectance, and total internal reflectance. The major advantage of reflectance sampling is that it does not suffer from the opacity problem as in the case of transmission sampling since the infrared light only needs to get reflected from the sample surface. Also the preparation time of the samples in reflectance sampling is the least and solid samples can be directly put under observation for the desired spectrum result.

Reflective sampling suffers from certain drawbacks as compared to the transmission sampling method. The first drawback is the involvement of a number of optical mirrors and lenses for focusing the infrared beam onto the sample and collecting the beam reflected back. These kinds of arrangements and accessories are expensive and can only be custom-made for a particular make and model of the FT-IR. In case of diffused reflectance, it is not possible to collect all the infrared light and the final obtained sample spectra ends up being noisy. Also the path length in reflectance sampling remains uncertain as the depth up to which the focused infrared radiation is able to penetrate is uncertain. This uncertainty in the path length hampers performing quantitative analysis of the sample as can be seen from the Beer's law. The depth of penetration generally is dependent on the angle of incidence, sample absorptivity, surface roughness, and sample geometry. Reflectance sampling cannot be applied for analyzing the bulk composition of the sample as the infrared light only gets reflected from the surface without penetrating into the sample. However, there is positive feedback of the sample surface characteristics and composition on the infrared spectrum obtained.

6.4 Data handling

The readings on the interferogram can be manipulated, adjusted, and identified for the desired results by applying mathematical and analytical procedures. These are apodization, resolution enhancement, phase correction, and mirror alignment.

6.4.1 Apodization

Apodization is a mathematical technique used to reduce the amplitudes of the side extremes adjacent to the absorption bands in the Fourier transform output of the original interferogram without degrading the desired level of resolution. Thus it can be noticed that with higher level of apodization, the resolution of the output tends to get reduced and vice versa and an optimal trade-off exists between the two. The term "apodization" comes from the Greek word "Arados," which means "without feet" and represents the sharp, low-intensity oscillations around the major spectral absorption band in the Fourier transformation of the interferogram. Two apodization functions are commonly used, which are the triangular and Happ-Genzel functions. These functions when used along with the interferogram eliminate the side oscillations around the absorption spectral band.

6.4.2 Resolution

Resolution is basically dependent on the maximum attainable optical path difference or retardation for a monochromatic source and it represents the smallest distinguishable range of spectral band that can be detected denoted by D_{max} cm^{-1}, where D_{max} represents the maximum attainable optical path. It is observed that rotational lines of water vapor can be appropriately resolved at a resolution of 2 cm^{-1} ($D_{max} = 0.5$ cm) while in contrast these lines are not visible at a lower resolution of 16 cm^{-1} ($D_{max} = 0.0625$ cm). Dynamic methods are used, which inherently enhance the resolution of the spectral bands due to the existence of different responses for highly overlapped bands.

6.4.3 Phase correction

Phase correction is a mathematical technique, performed to nullify the sine component of the interferogram (which is supposed to give only the cosine components) caused due to nonideality of the beam splitter. A double-sided interferogram that gives the correct power spectrum without affecting the magnitude can be used to avoid the ambiguity. On the other hand, in a single-sided interferogram the ambiguity exists and phase correction is required. Two of the commonly used mathematical techniques in phase correction of a single-sided interferogram include Mertz algorithm and Forman algorithm. In the Mertz algorithm, the largest data points are assigned a "zero" retardation point and the amplitude spectrum is measured with respect to that point. The phase array of the double- sided interferogram can be then used to phase-correct the power signal of the transformed interferogram. Forman algorithm is very much similar to the Mertz algorithm with all the measurements performed in the retardation space.

6.4.4 Mirror misalignment

Mirror misalignment is a problem in the interferometer, which can cause lowering of the intensity of a scanned single beam versus wave number graph. In continuous scan FT-IR, the accuracy of the movable mirror speed is sensed by the laser fringe counter as a constant scan velocity of the mirror is desired in it. Correction signals are generated when a misalignment occurs, which directs the mirror to move at a constant speed. In step-scan FT-IR, data are collected keeping the optical path difference constant or oscillating about a fixed point. The collected data is a function of time and the intensity of the infrared radiation is modulated either by amplitude modulation or by phase modulation.

6.5 Data interpretation

The analysis of the sample absorption spectrum obtained in an FT-IR spectroscopy involving transmission sampling is based on Beer's law, which gives the relation between the concentration of a molecular species in a sample and the absorbance spectrum. When light beam passes through a thin film of the sample, a few of the photon particles are absorbed by electrons in a molecule (that varies with different material) to reach a higher energy level. It is due to this absorption phenomenon that the photons received on the detector are reduced and hence a reduction in intensity of the light beam. According to Beer's law, absorption of a light beam at a specific wavelength by a material species is mainly dependent on the product of the path length and the concentration of the absorbing material species in the sample. This can be given as below:

$$A = \varepsilon L C$$

where A is the absorbance, L is the path length through the sample, C is the concentration of the absorbing species, and ε is the absorptivity of the species. Absorptivity is the fundamental property of a pure molecule for a fixed wavelength of light beam just as any other properties of material such as boiling point, molecular weight, etc. However for a single species, absorptivity varies with change in wave numbers or frequency. Again, at the same wave number two different materials can never show similar absorptivity. Since absorbance is a unitless quantity, absorptivity has the unit of (length × concentration)$^{-1}$.

In order to apply the Beer's law in the quantification analysis of an unknown sample, two of the major steps that are needed to be followed are calibration of the unknown sample with the standard sample and prediction of the properties of the sample. While conducting the FT-IR spectroscopy of a sample, it is assumed that the calibration applied is the same in both the unknown and standard sample.

6.5.1 Calibration

Calibration is done using the Beer's law, the equation of which is analogous to a straight line, $y = mx + b$, where m is the slope.

From the above Beer's equation, $m = \varepsilon L$.

A calibration line is plotted between the peak height of the absorbance spectrum at a specific wave number of standard samples and their corresponding concentrations (measured in masses and volumes). The concentrations are to be obtained by methods other than FT-IR. A straight calibration line indicating a majority of points passing through it represents a chemical system that supports Beer's law and is often a case for many sample experiments. A few samples however fail to obey Beer's law with a positive intercept on the y-axis (absorbance axis) even when the concentration is zero. This is due to presence of noise and influence of other chemical species in the formation of absorption spectrum. After generation of the calibration line, coefficient of correlation of the line is obtained, which usually is the value 1 or close to 1 (0.999 etc.), indicating a strong positive correlation between the absorbance and the concentration.

6.5.2 Prediction

After calibration, the next step is to predict the concentration of the infrared absorbing species that is being placed in the infrared beam using the absorption spectra obtained. The Beer's law is rewritten for an unknown sample, the slope being the same as obtained during calibration with the standard samples:

$$C_{un} = \frac{A_{un}}{m} \tag{6.2}$$

where C_{un} and A_{un} are concentration and measured absorbance of the light-absorbing species in the unknown sample. The assumption that the slope "m" is constant for both the unknown and the standard sample requires controlling of a number of experimental variables in the unknown sample (affecting the absorbance peak height) to perform a successful quantitative spectroscopic analysis.

6.5.3 Measurement of absorbance

Correctly measuring the absorbance of the sample can be a difficult task due to the presence of other substances in the sample, which gives an overlapping absorbance spectrum. In order to tackle this, the absorbance spectrum of the other substances must first be measured separately and then compared with the absorbance of the sample. For measurement of the absorbance, generally the peak heights of the spectrum are measured, which might sometimes drift up and down in the y-axis due to various unwanted problems in the sample and the surrounding. It is due to this, a baseline is constructed that joins the initial and the final position of the absorbance peak. This is a line considered as "zero" absorbance and all other absorbance peaks are measured with respect to this baseline. The peak area is measured by integrating the area of the peak from the beginning till the end position of the baseline.

6.6 FT-IR spectroscopy: Nanofillers and nanocomposites

In the present section of the chapter, a variety of nanofillers such as carbonaceous nanofiller, clay, metal oxide nanoparticles, and their nanocomposites have been explored by the FT-IR spectroscopy.

6.6.1 Identification of metal oxide nanofillers

An absorption peak is observed below a wavelength of 1400 cm^{-1} for metallic nanoparticles from oxides and hydroxides. This region is usually referred to as the fingerprint region. For Fe_3O_4 nanoparticles, Tab. 6.1 provides the details for the characteristic peaks of Fe_3O_4, γ-Fe_2O_3, and α-Fe_2O_3.

Tab. 6.1: Band assignment for iron oxide in the FT-IR spectra.

Iron oxide	Assigned vibrations (cm^{-1})
Fe_3O_4	600, 440
α-Fe_2O_3	537, 458
γ-Fe_2O_3	639, 588

Figure 6.9 depicts the FT-IR spectra obtained for the ZnO nanoparticles in a KBr matrix. A broad band corresponding to the vibration of the –OH group of water is reflected in the wavelength at 3392 cm^{-1}. This indicates that water has been absorbed at the surface of the ZnO nanoparticles. The bending of –OH group is reflected in the band at 1608 cm^{-1}. The stretching of Zn-O is attributed to the strong band at 465 cm^{-1}.

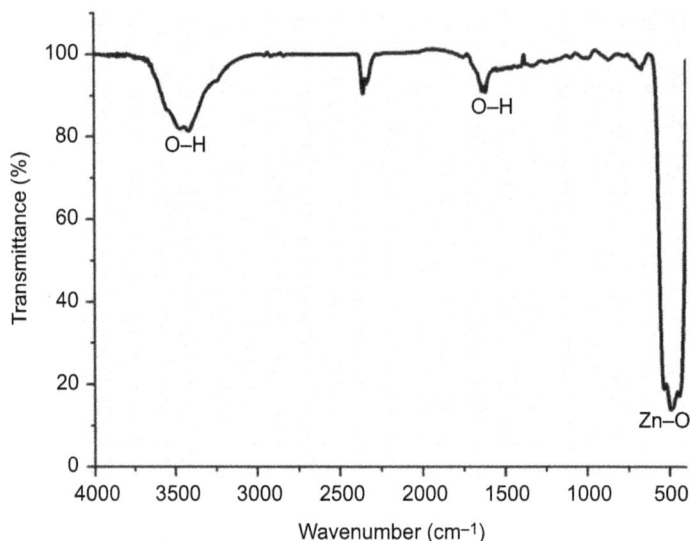

Fig. 6.9: FT-IR spectra in case of ZnO nanoparticles (reproduced with permission from [24]).

Similarly the spectra for γ-MnO$_2$ have been delineated in Fig. 6.10. The stretching of the –OH group is reflected in the band at 3439 cm^{-1}. The bending of –OH group together with the Mn atoms is reflected in the bands at 1644 cm^{-1}, 1430 cm^{-1}, and 1136 cm^{-1}. The vibrations of Mn-O are reflected in the bands at 560 cm^{-1} and 530 cm^{-1}.

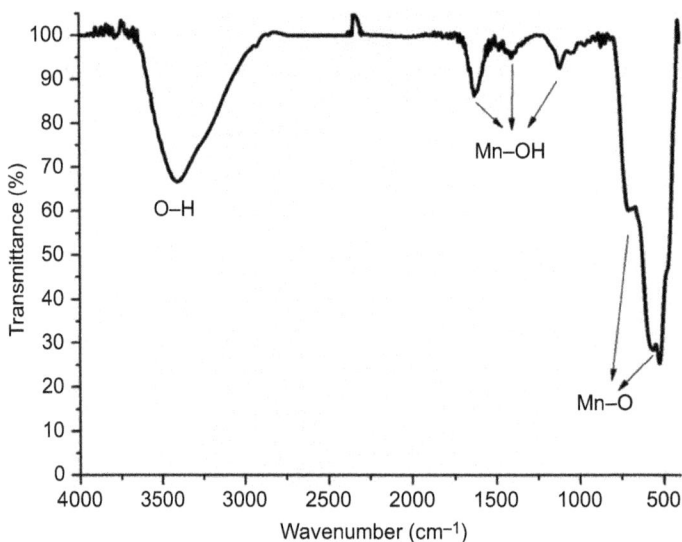

Fig. 6.10: FT-IR spectra in the case of MnO nanoparticles (reproduced with permission from [24]).

As such the results of the obtained FT-IR spectra reveal the presence of water in the MnO_2 structure.

Figure 6.11 shows the FT-IR spectra for the titanium dioxide (TiO_2) nanoparticles synthesized through the sol–gel process. The stretching of the –OH group is reflected in the broad band between 3800 cm^{-1} and 3000 cm^{-1}. This signifies the presence of water as moisture. Stretching of titanium carboxylate is attributed to the peak at 1635 cm^{-1}. The stretching of the peaks between 800 cm^{-1} and 450 cm^{-1} reflects the stretching of the Ti-O bonds.

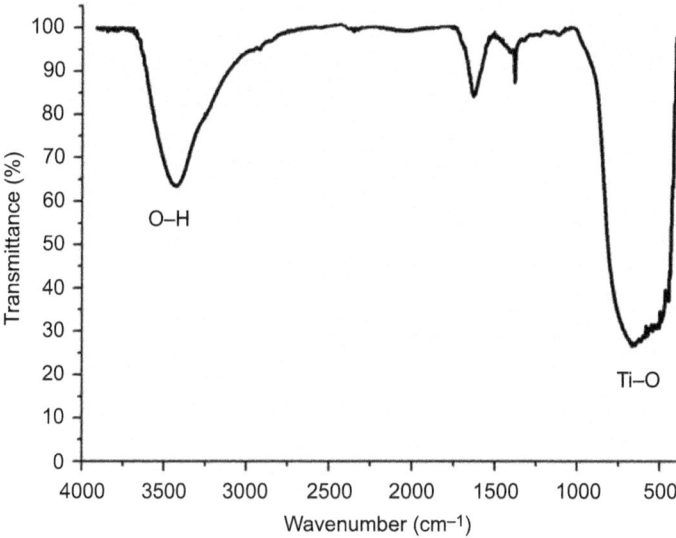

Fig. 6.11: FT-IR spectra in case of TiO_2 nanoparticles (reproduced with permission from [24]).

6.6.2 Identification of carbonaceous nanofillers

The industrially produced single-walled carbon nanotubes and multiwalled nanotubes have been investigated by the FT-IR spectroscopy [25]. The same vibrational modes are exhibited by the two single-walled nanotubes in the wavelength ranging from 700 cm^{-1}–2500 cm^{-1}. However, a shift in energy has been observed with different intensities. A shift of 25 cm^{-1}, 20 cm^{-1}, and 18 cm^{-1}, respectively, in the peaks at 1110 cm^{-1}, 1535 cm^{-1}, and 1700 cm^{-1} has been observed in the CNTs-ILJIN sample. In both the cases, the low shoulder is observed between 820 cm^{-1} and 840 cm^{-1}. The wavelengths of 868 cm^{-1} and 1590 cm^{-1} represent the characteristic modes for graphite. Strong structures near the vicinity of graphite modes have been observed in the FT-IR spectra.

The researchers investigated the FT-IR spectra for graphite, graphene oxide, and reduced graphene oxide [25]. No significant peaks were observed in the case of the

FT-IR spectra for graphite. Table 6.2 lists the presence of different functionalities in graphene oxide obtained through the FT-IR spectroscopy.

Tab. 6.2: Band assignment for iron oxide in the FT-IR spectra.

	Assigned vibrations (cm^{-1})
O-H stretching	3400
C-O stretching	1720, 1060
C-OH stretching	1220

The vibrations associated with the stretching of O-H in case of the reduced graphene oxide has been revealed to be at 3400 cm^{-1}. This was reduced significantly because of deoxygenation. The stretching vibrations for C-O were still observed at 1720 cm^{-1}. Sharpness in C-O vibrations at 1060 cm^{-1} was revealed because of the hydrazine reduction.

6.6.3 Identification in GO/ZnO nanocomposites

The FT-IR spectra for GO/ZnO nanocomposites have been depicted in Fig. 6.12. The peaks corresponding to O-H, C-OH, and C-C are reflected in peaks at 3431 cm^{-1}, 1211 cm^{-1}, and 1582 cm^{-1}, respectively. The peak at 431 cm^{-1} on the other hand reflects the Zn-O bonding.

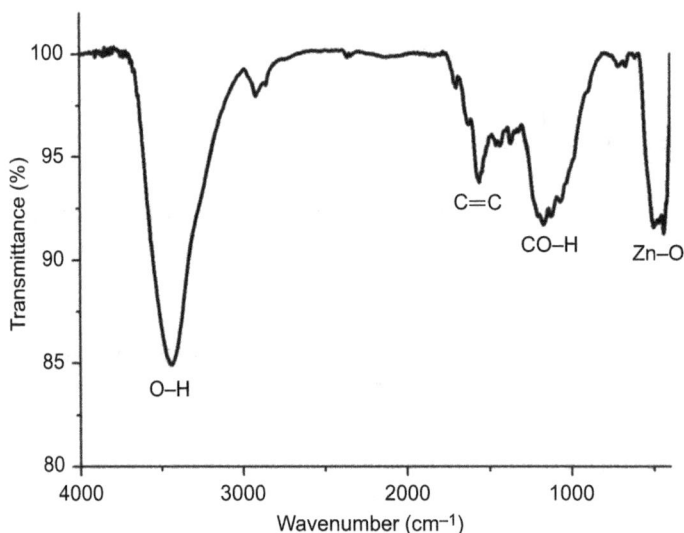

Fig. 6.12: FT-IR spectra in case of GO/ZnO nanocomposites (reproduced with permission from [25]).

It is worth to be noted here, that the FT-IR spectra do not reveal the intermetallic bonds. For instance, in the FT-IR spectra (Fig. 6.13) of GO/Au nanocomposites, only the peaks corresponding to GO are observed. The identification of intermetallic bonds is revealed in the Raman spectrum obtained through Raman spectroscopy. Other characterization techniques such as XRD and UV spectroscopy can also be used to characterize such nanocomposites.

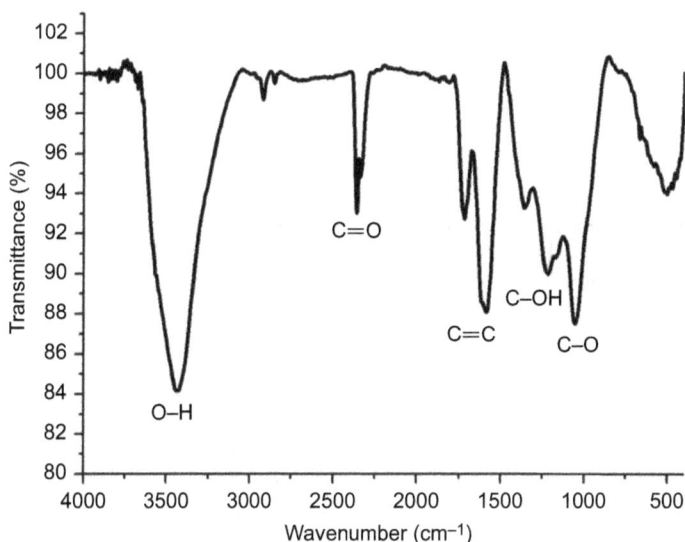

Fig. 6.13: FT-IR spectra in the case of GO/Au nanocomposites (reproduced with permission from [26]).

The FT-IR spectra for granular activated carbon/TiO_2 nanocomposites are revealed in Fig. 6.14. The C-O-C, C-O, and C-C bonds can be seen in the peak bands of 1200–1700 cm^{-1} and 1000–1200 cm^{-1}. The strong peaks between 3200 cm^{-1} and 3600 cm^{-1} reflect the spectrum corresponding to TiO_2. The peaks corresponding to the aforementioned range reflect upon the stretching vibrations of the bond between Ti and hydroxyl group.

The FT-IR spectra for GO/Fe_3O_4 are delineated in Fig. 6.15. The band at 588 cm^{-1} reflects on the vibration corresponding to the Fe-O bonds that are present in Fe_3O_4. The peaks corresponding to GO have been specified in the spectrum depicted in Fig. 6.15.

6.6.4 Identification in case of polymeric nanocomposites

Polymeric materials have been playing a quintessential role in various domains of modern life and hence an understanding of the associated properties and the chemi-

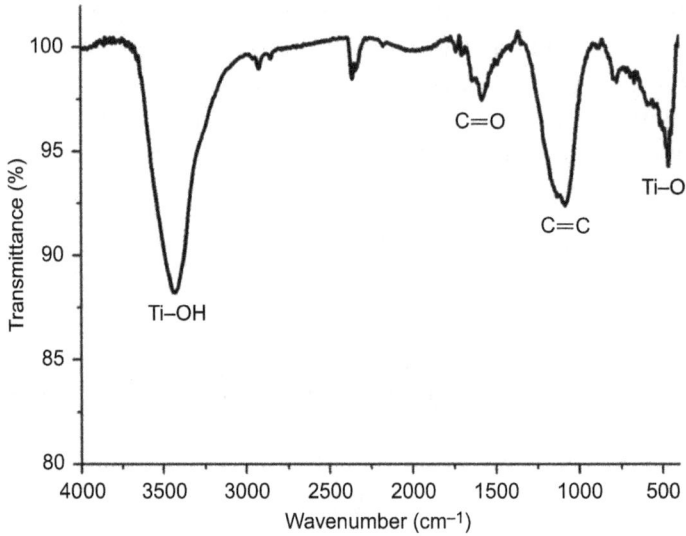

Fig. 6.14: FT-IR spectra in the case of GAC/TiO$_2$ nanocomposites (reproduced with permission from [27]).

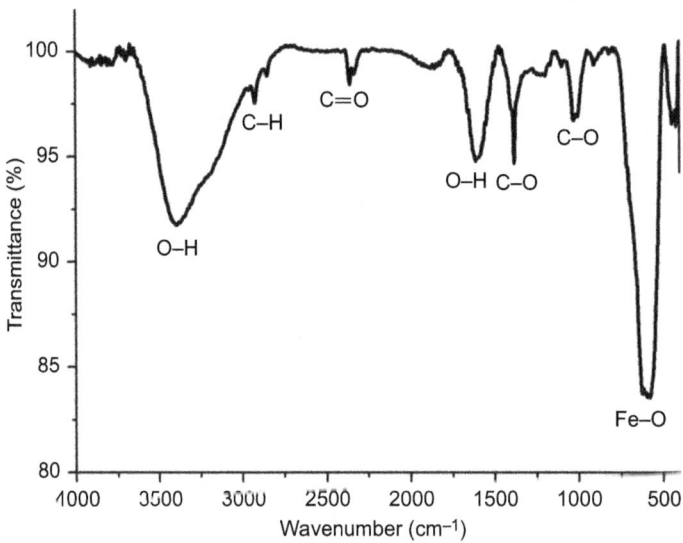

Fig. 6.15: FT-IR spectra in the case of GO/Fe$_3$O$_4$ nanocomposites (reproduced with permission from [28]).

cal structure is equally important. Vibrational spectroscopy is really advantageous in deciphering the probable polymeric structure and is employed extensively for all kinds of polymeric structure. The IR spectroscopy method is convenient in case of such materials and is advantageous over the chemical analysis in the sense that it avoids the exposure to the toxic and corrosive chemicals. In the case of polymeric nanocomposites, the IR spectrum will be the sum of the spectra of the individual constituents in case the components are immiscible. On the other hand in case the components are miscible, then there are evident chemical interactions between the constituting ingredients. This is reflected in the shifts in the wave numbers and the band broadening. Here in the present context, spectra of a few polymeric nanocomposites have been discussed as example. For instance, consider the FT-IR spectra of polystyrene/TiO_2 nanocomposites. The main bands associated with polystyrene are five peaks: the range 2800–3100 cm^{-1} reflects on the C-H stretching, the deformation and the vibrations of C-H are reflected in peaks at 1601, 1492, 1451, 1027, 757, and 698 cm^{-1}. The same IR spectra can be observed in pure polystyrene film and its nanocomposites with small percentages of TiO_2. However, characteristic peaks for TiO_2 can be observed in polystyrene nanocomposites with higher percentages of TiO_2. The vibrations in the aromatic rings are reflected in the spectral bands at 3081, 3060, 3026, 1602, and 1493 cm^{-1}. The other peaks at 2922, 2851, and 1452 cm^{-1} are due to the aliphatic backbone of PS.

Figure 6.16 reveals the FT-IR spectrum for polycarbonate and its nanocomposites with TiO_2. Table 6.3 delineates the band assignment for the different functionalities for polycarbonate.

Tab. 6.3: Band assignment for polycarbonate and polycarbonate/TiO_2 nanocomposites in the FT-IR spectra.

	Assigned vibrations (cm^{-1})
Free alcohols	3684
C-H stretching	3059
Asymmetric stretching mode of CO_2	2411, 2052
Phenyl ring stretching	1604
Skeletal vibration of phenyl compounds	1512
Bending mode of CO	852

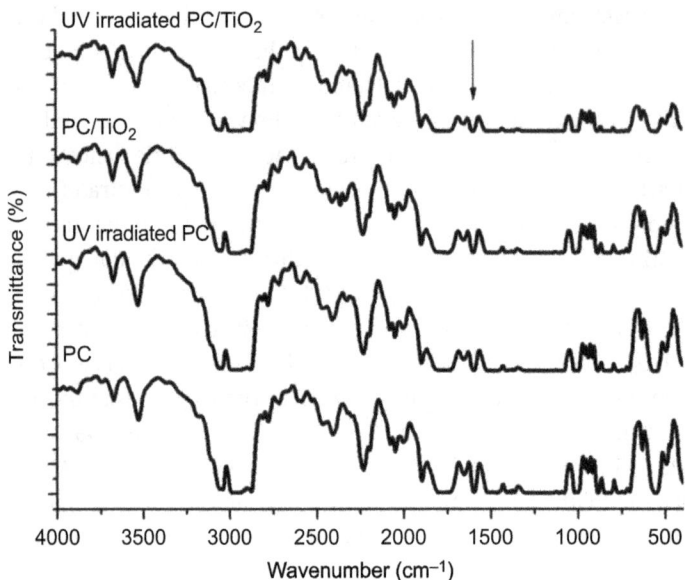

Fig. 6.16: FT-IR spectra in case of PC/TiO$_2$ nanocomposites (reproduced with permission from [29]).

6.7 Conclusion

IR spectroscopy has been considered to be one of the versatile methods to investigate the chemical structure of any material. The structure of the molecules within the material can be assessed through the IR spectrum delineating the absorption or transmittance with respect to the IR frequency. The absorbance of the IR radiation on passing through the sample takes place in the form of vibrations, bending, and stretching. FT-IR is however a more preferred technique over IR. Higher speed and higher signal-to-noise ratios are some of the advantages of the FT-IR instruments over the other existing dispersive instruments. FT-IR has been acknowledged by the scientific community as one of the best methods to decipher several aspects associated with the molecular structure.

Chapter 7
Raman spectrometer

7.1 Introduction

The phenomenon of Raman spectroscopy generally was first discovered by Sir Chandra-shekhar Venkata Raman in 1928 using sunlight as the source of light, a telescope as a collector, and his eyes as a detector. With further development in new technologies such as a variety of laser excitation sources, a variety of improved detectors, collimators, lenses, polarizer, monochromators, etc., the sensitivity of Raman spectroscopy has enhanced tremendously and minute details of the spectrum exhibiting the nature of the sample can be extracted.

In the initial stage, development was mainly limited to the improvements in excitation sources for achieving a highly monochromatic input light. Various lamps such as helium, lead, zinc, and bismuth lamps were developed but due to their low intensities they were not considered fit for a variety of studies of the samples. Mercury excitation sources were first used for Raman spectroscopy in the 1930s. Later a mercury excitation source system consisting of four mercury lamps surrounding the sample was developed for Raman instruments by Hilger Co. Ham and Walsh [30] developed microwave-powered lamps of helium, sodium, potassium, rubidium, and mercury. The use of these elements including argon, cesium, and rubidium as an excitation source for colored materials was later also examined by Stammreich [31–35]. Ar$^+$ and Kr$^+$ lasers were later developed and credited with their suitability in Raman spectroscopy, and recently Nd-YAG lasers have been used extensively in Raman spectroscopy.

In the detection system, photographic plates were first used, which suffered from the cumbersome and time- consuming production of these plates. Later in the year 1942, the photoelectric Raman instrument was developed by Rank and Weigand [36] using cooled cascade-type RCA IP21 detectors followed by cooled RCA C-7073B photo-multiplier in 1950.

During the early 1960s, developments were made in the optical train system where observations have been made regarding the reduction of stray lights affecting the Raman spectrum by using double and triple monochromators instead of using just the single one. The efficiency of collection of the scattered spectrum was further achieved by the introduction of holographic gratings.

Raman spectroscopy involves a number of doubt-clearing questions that are dependent on the availability of instruments, type of result that is desired, and sample forms that are used before even conducting experiment on it, which need to be answered. Questions like which laser exciting source needs to be used (a UV laser source or near-infrared or visible laser sources), what type of detector such as charge-coupled device (CCD), indium gallium arsenide (InGaAs) or Fourier transform (FT) interferometer are to be used (based on the desired accuracy in results that is required),

https://doi.org/10.1515/9783110997590-007

whether detectors need to be used with dispersive monochromator, and what accessories are available to conduct the Raman experiment, are important to be known by the spectroscopist. Other questions might arise from the setting of the instruments for achieving the desired spectrum such as: how to place the sample at the focus of the incoming laser excitation source so that degradation, burning, or fluorescence can be avoided or how these settings might affect the sample result of the Raman spectrum A thorough research on all these questions needs to be conducted and better optimized settings must be adopted for the desired quantitative and qualitative interpretations.

7.2 Instrumental detail, working, and handling of the Raman spectroscopy

Raman spectroscopy mainly consists of components (Fig. 7.1), such as excitation source, sample illumination, and collection system, wavelength selector, and detection and processing system [37]. All the accessory instruments are selected and are influenced by the wavelengths of the instrument's excitation source, detection range, and the type of wavelength selector.

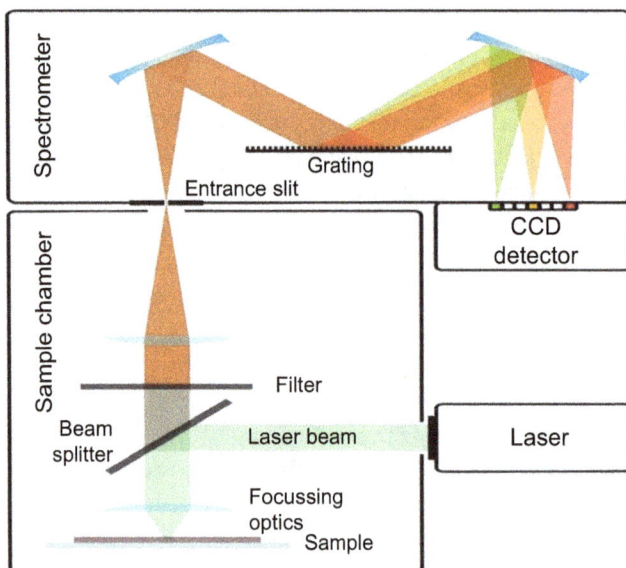

Fig. 7.1: Components of a Raman spectrometer.
Source: Setup Raman Spectroscopy (2019), Wikimedia commons (reproduced with permission from [38]).

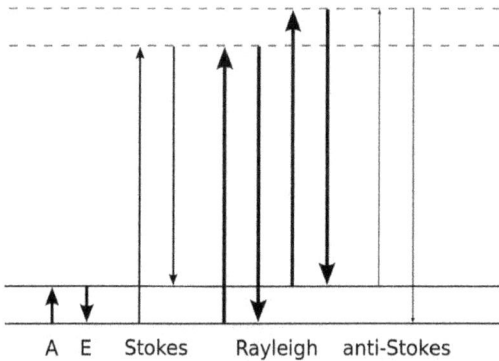

A E   Stokes   Rayleigh   anti-Stokes

Fig. 7.2: Raman and Rayleigh spectral lines.
Source: Raman scattering (2007), Wikimedia commons (reproduced with permission from [39]).

7.2.1 Excitation source

Mercury lamps and filters (for a fixed range of wavelength) and high power light sources were mostly used before the introduction of laser source in the 1960s. But due to higher delay in warming and igniting on these lamps, laser sources were preferred and adopted by many researchers and spectrographic community to detect Raman scattering.

Commonly used laser sources for Raman spectroscopy includes continuous wave lasers (Ar^+, Kr^+, and He-Ne), pulsed lasers (Nd:YAG), and excimer lasers. The advantages of laser over other excitation sources such as Mercury lamp and filters are as follows:

1. Laser light sources produce powerful beams with pulsed lasers having peak power in the order of 10–100 MW.
2. Laser light sources are highly monochromatic where the extraneous light is much weaker and can be eliminated using notch filters and pre-monochromator.
3. Laser beams are highly concentrated having a smaller diameter that delivers high power on a small section of the sample and facilitating studies on small crystals and microliquids.
4. Laser sources are polarized and are employed in measuring depolarization ratios.
5. Using dye lasers and devices, laser sources can be obtained in a wide range of wavelengths.

7.2.2 Continuous wave lasers

Continuous wave lasers (CW lasers) are generally gas lasers, which emit light in the visible spectrum and involve passing of high voltage current through noble gases like Ar and Kr contained in a plasma tube. As the discharge ionizes the noble gas, the tube is constantly cooled by a water jacket surrounding it and the light that is emitted is directed through the Brewster window attached to the two ends of the tube for a polarized output beam. The window is having a Brewster angle of 55.6°, which is basically dependent on the refractive index of the window material. The CW lasers are generally operated at a higher wavelength in the range of around 325–800 nm. The 488 nm line of argon is about 2.8 times higher intense than the 632.8 nm line of He-Ne. Plasma lines can cause problems due to interference formation and are therefore optically filtered. These plasma lines can be characterized by their high bright and sharp lines that shift on changing the frequency of the exciting line. Interference filers were generally used but suffer from the drawback that filtering must be done for each wavelength of the electromagnetic spectrum. Notch filters are an improved version of the interference patterns that deliver a narrow bandwidth of the spectrum.

7.2.3 Neodymium-YAGs

Neodymium-YAGs (Nd-YAG) are solid state lasers where the YAG (yttrium aluminum – garnet) remains in rod shape and host the Nd^{3+} ions, which do the lasing. Nd-YAG has a lasing wavelength of 1064 nm (with outputs that can exceed 10 W), which can be reduced to half and outputs ranging in 50–200 mW with the help of an intracavity crystal. Nd-YAG is ideal for use as an excitation source in FT-Raman spectroscopy.

7.2.4 Diode lasers

Diode lasers generally use the excitation at the p-n junction of a semiconductor producing light at a specific wavelength ranging in between the blue and the infrared regions of the spectrum, though developments have been made to reduce the wavelength to UV region. Their small size, efficiency, minimum power, and cooling requirement are some of the advantages of diode lasers and are extensively used for various applications such as in CD readers/writers and fiber-optic communications. For maximizing the intensity of a particular wavelength light and controlling the tendency to drift in its wavelengths, improvements have been made to keep the diode at a temperature not varying beyond ±0.01 °Cs and by providing grating in the laser cavity.

7.3 Sample illumination and collection

Sample illumination is done with a laser beam of smaller diameter (1 mm) being directed on the surface of the sample, the scattered light of which is collected at an optical orientation of 90° or 180°. The illumination and collection system are important as the Raman scattering is weak and must be efficiently collected by an achromatic lens system consisting of a collector lens and a focusing lens. In oblique illumination, the laser beam is directed at an angle close to 85° from the normal to the surface of the sample and the use of collecting lens is avoided and instead an ellipsoidal mirror is used to focus the scattered UV on the entrance slit. Achromatic lens system is necessary as it eliminates the need to adjust the position of the observing lens and its focus to obtain Raman signal and need not repeat the procedure with changing laser wavelengths. While making adjustments, one should take into consideration the hazardous effect of the laser radiation, the laser power must be put to minimum, and utmost care should be taken to protect the eyes.

The light collection power of a lens can be expressed in terms of the "F" number by the formula given below:

$$F = \frac{f}{D} \tag{7.1}$$

where "f" represents the focal length of lens and "D" is the diameter of the lens. For a larger collecting power the "F" number must be smaller or (by decreasing the focal length and increasing the diameter of the lens). A bright image might sometimes arise in the case of 90° and the oblique scattering configuration due to fluorescence or reflection of the light from reflective surfaces of the glass and the sample. Therefore, caution must be exercised and image processing of the irradiated sample at the entrance slit of the monochromator must be done with proper sequential procedure.

7.3.1 Wavelength selector

Interference filters are the simplest and the most common type of wavelength selectors, which allow selective wavelength of light to pass through a flat semi reflecting medium of dielectric that has thickness of half the desired wavelength of light. Variable wedge-shaped interference filters are used commonly but suffer from the drawback that the spectral resolution is too large for Raman spectra.

Though monochromators are commonly used to analyze Raman spectrum developments in FT Raman have given tough competition to the use of monochromators in Raman spectroscopy. These are detailed below.

7.3.2 Monochromators

Monochromators split the incident light into its constituent wavelengths by making it fall on the face of the grating, which along with the undiffracted light causes stray lights, blocking the visibility of the Raman spectral lines (Fig. 7.2). These stray lights can be reduced by passing the output light of the first monochromator through the second monochromator and further by using the third monochromator. This enables sharp visualization of the Raman spectrum and Raman lines are close to the Rayleigh lines. Resolution is another important factor in obtaining Raman spectral line, which is affected by the slit width, increment rate in data points, and the grating. Higher resolution is obtained with reduced slit width and thus lowering the power of the laser light (P) that is related to the slit width as given by: $P = P_0 \times SW^2$ where P_0 is the power of the incident light and SW is the slit width. With higher increment speed (scanning speed of the spectrum) the resolution of the spectrum is reduced, which can be avoided by moving the monochromator in smaller increments. In terms of gratings, with increased grooves per millimeter though the resolution of the spectrum increases, it is accompanied by larger dispersion and loss of power. This loss of power can generally be compensated by widening the width of the slit. Mechanical backlash in the measurement of the spectrum on instruments and temperature effect, which basically shift the band position by over 3 cm^{-1}, are other factors that need to be controlled during the operation.

7.4 FT Raman spectroscopy

FT Raman spectroscopy is an integrated system between the conventional Raman sampling module and the FT-IR system, which enables it to operate at near-infrared region, thus eliminating or reducing the fluorescence effect. It basically consists of a laser source, the radiation of which is directed on to the sample by means of a mirror and lens. The scattered radiation from the sample is then collected and using a parabolic collector to the beam splitter which splits it into two, directs it to the moving and the fixed mirrors and passes on to the interferometer head. The radiation is passed through a series of dielectric filter before focusing it to a liquid nitrogen-cooled detector (Ge).

Higher fluorescence makes it difficult to obtain the Raman spectral lines due to noise. Nd:YAG lasers are generally used as an excitation source for the FT Raman spectroscopy operating at a wavelength of 1064 nm. Various FT-IR capabilities including computer programs and manipulations can be made use for detecting Raman spectral lines. Apart from the fluorescence problem, conventional Raman spectroscopy suffers from low resolution and difficulty in precisely detecting frequencies. Improvements in FT Raman technology had enabled eliminating these drawbacks in conventional Raman spectroscopy. FT Raman spectroscopy is commonly termed as time domain spectroscopy because unlike the conventional Raman spectroscopy where the intensity of the laser light is plotted against the frequency or the wave number, FT Raman is used to plot intensities of different frequencies simultaneously.

FT Raman has the disadvantage of reduction in the power of the laser source (10 W) to about 1 W of useful power, which is critical to all the optical components in the instrument. Filters must also be equipped in order to eliminate stray lights from the excitation of the laser source as it can saturate the electronics and detectors. Filters are often used to reduce the intensity of the Rayleigh lines (which are usually 10^6 times the intensity of the stoke lines) and make it comparable to the strongest Raman lines. Holographic notch filters are an effective means to reduce the intensity of noise at the laser frequency making it free from the Raman lines. Plasma emission filters are commonly used in order to eliminate the optical output of the He-Ne laser so that interference with the main laser light can be avoided due to high sensitivity of the detectors to the He-Ne laser wavelength.

Although FT Raman is advantageous in many respects as compared to the conventional Raman spectroscopy, a few special cases where FT Raman technique fails to provide suitable results can be realized. A few materials that absorb wavelength in the IR region such as charge transfer conductors, polycyclic aromatic compounds, and transition metal complexes have the problem of fluorescence in detecting Raman spectrum. FT Raman will not detect impurities in parts per million through spectral subtractions and will not disperse the visible laser Raman spectrum. Due to blackbody radiation of the sample at a temperature greater than 150 °C, which basically makes the background intense, making it almost impossible to detect the Raman spectrum, FT Raman cannot be used with Nd-YAG laser.

7.4.1 Detection of the Raman spectrum

Detection and amplification of Raman spectral lines was very difficult during the early period of development and was accomplished by micro photometers examining the long-exposed photographic plates. With the introduction of advanced laser source for monochromatic light and sensitive detectors, a number of techniques have been devised in order to detect and amplify the Raman spectral lines. The major ones have been described below:

7.4.1.1 Photon counting

In a monochromatic and single detector system, photomultiplier (Fig. 7.3) is generally used to collect the scattered photons at the focal plane of the sample, which consists of a photocathode, a series of dynodes, and an anode along a tube. When photons strike the photocathode, electrons get released and pass through the dynodes with lesser negativity than the cathode but higher negativity than the anode. After each layer of dynodes, the electrons are multiplied in number and when they finally reach the anode, they receive more number of electrons than previously, which amplifies the signal. The quantum efficiency or the sensitivity of the photocathode generally tends to be at the maximum with lesser wavelength and reduce with higher wavelength of the collected spectrum.

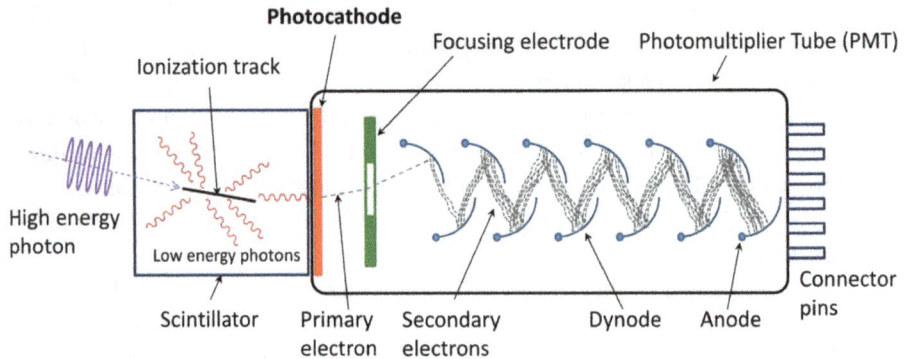

Fig. 7.3: Photomultiplier.
Source: Photomultiplier tube and Scintillator (2013), Wikipedia (reproduced with permission from [40]).

The electrons collected at the anode are averaged over time for processing using DC amplification or by a separate photon counting method, though photon counting method gives a higher sensitivity than the former one.

7.4.1.2 Photodiode array

Generally, the scattered photons collected on the detector are able to scan a particular frequency of Raman spectral lines at an instant in normal Raman spectroscopy but this process is not suitable for measuring Raman lines in unstable compounds. To avoid this, multiple photo detectors are arranged in an array (Fig. 7.4) to detect photons at different frequency ranges simultaneously and convert the optical output into charge signals allowing detection of changing Raman lines in unstable compounds. Photodiode arrays are not as sensitive as that of the photon counting method.

7.4.1.3 Charge coupled device (CCD)

CCDs are commonly used in modern Raman spectroscopy, which is basically an array of photosensitive elements that directly receives the photons from the spectrometer devices to release electrons and convert them into small charges. CCDs are generally made out of silicon- based semiconductors arranged in the abovementioned array. The charges stored are read by an analog to digital convertor, one element at a time, for which the array element charges are transferred in the left to right direction, while

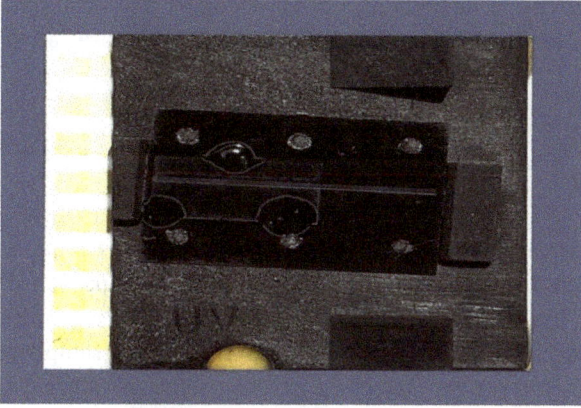

Fig. 7.4: Photo diode array.
Source: Photodiode array chip (2013), Wikimedia commons (reproduced with permission from [41]).

the charges in each row are transferred from bottom to top. Currently employed CCD (Fig. 7.5) has the format of 1024 × 1024 pixels with each pixel having the size of 25 mm^2.

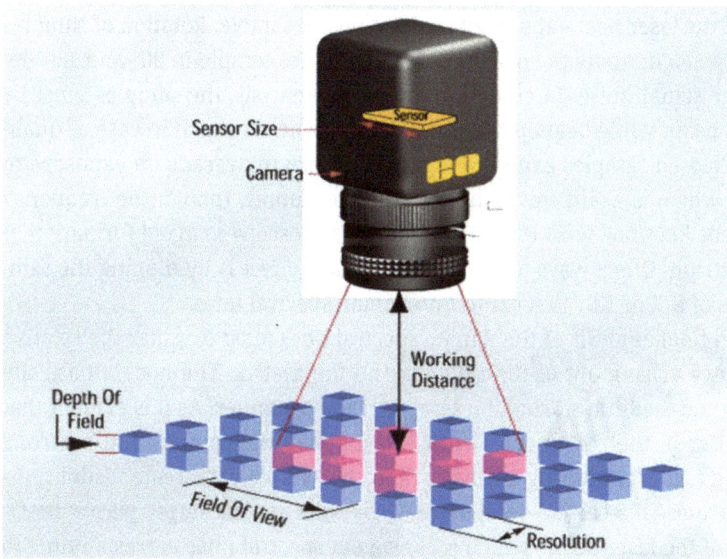

Fig. 7.5: Charge-coupled device (CCD).
Source: Diagram of a charge coupled device (2004), Wikimedia commons (reproduced with permission from [42]).

The advantage of employing CCDs as photo detectors is its low readout noise levels, which completely eliminates the requirement of optical intensification, thus enhancing quantum efficiency and sensitivity to higher range of wavelengths (120–1000 nm). Raman spectrum in fluorescent compounds that emit light at near-IR range are easy to detect using CCD's.

7.5 Sample preparation

Sample preparation is relatively easy in the case of Raman spectroscopy as compared to other spectroscopy techniques (infrared spectroscopy) [43, 44]. The sample can be in the form of powders, liquid, irregular or crystal solids, vapor, and polymers, which can be mounted using a number of techniques or by simply exposing the sample to the laser focus. Various liquid-holding tubes, clamps for irregular solids, powder holders, and cuvette holders are used for mounting the sample on to the laser beam. A few other advanced mounting techniques involve specialist cells for rotating sample, variable temperature and pressure cells, reaction cells, and vapor cells.

Accessories for sample mounting in Raman spectroscopy come in a variety of physical forms and temperature. Powdered and liquid samples are commonly used for lab and industry purposes, which can be directly fed in the path of the incoming laser beam. Solid samples having higher fluorescence are usually rotated to avoid the burning effect of the laser beam at a fixed position on the sample. Rotation of samples in the beam can also eliminate the orientation effect of the sample in 90° and 180° by averaging out the signal noise. In case of quantitative analysis, the samples must be optimized in the beam while beam positioning does not matter much in case of qualitative analysis. Colored samples exhibit a higher tendency to degrade on exposure to incoming beam, which is again solved by rotating the sample, though the frequency of rotation must be kept below 50 Hz in FT Raman spectroscopy to avoid formation of beats in the spectrum. Other ways to avoid fluorescence effect is by dipping the sample into solutions of KCl or KBr that exhibit no Raman spectral lines.

The recorded final analysis of the Raman spectral lines must consider the relative intensity difference arising out of the samples and the matrix. The matrix basically represents the liquid medium storing the sample or the container. As it is evident that spectral lines differ in their intensities with varying material composition, a strong material emitting higher scattering intensity than the sample can create visibility issues of the spectrum. This can be demonstrated by selecting an empty plastic bottle kept in the path of the laser beam, which will give out spectral lines corresponding to the polythene. But the phenomenon alters when sulfuric acid is poured into the plastic container and placed in front of the beam. The spectral line exhibiting now will correspond to that of the sulfuric acid. Water and glass exhibit weak Raman spectral lines that do not affect the sample spectrum and therefore are commonly used for Raman spectroscopy purposes. For smaller samples or even for checking the homogeneity of larger samples, microscopes or microprobes are generally used, when the need to constrict the diameter of the laser beam arises.

7.6 Sample handling

Researchers and authors have long been experimenting with samples being placed inside plastic containers, vial, bottles, and brown bottles and radiated with laser beams. It was found that the clarity of the Raman scattering was usually unaffected by any medium between the sample and the laser source if the sample spectrum is strong, provided the outer surface of the container and the sample remain clean and free from fingerprints. The above statement does not hold true with samples having weak Raman spectral lines where the spectral lines scattered out from the container dominate. Other factors affecting the sample's Raman spectral lines that do not apply to the above statement are fluorescence and burning. A few methods and techniques which enhance the Raman spectrum and diminish the factor effects and interferences were developed. One of these techniques involves the placement of the sample in such a way that the laser beam is focused into the sample away from the walls of the plastic cuvette container or from the sides and foot of the microtiter plates. This en-hances the power density of the scattered emission from the sample, mitigating the interference effect of the plastic cuvette and microtiter plates. On the other hand, if the laser beam is focused onto the surface of the plastic cuvette, the spectral lines de-tected mostly belong to that of the polymer rather than the spectrum of the sample.

Crystalline solid samples have been mounted in compacted holders by research-ers, the spectral radiation of which was found to be dependent on the orientation and particle sizes. Neat powders stored in loose containers have also been illuminated by laser beams and found to get displaced by the power of the beam, exhibiting a poor Raman spectrum. This can be tackled however by fixing the particle (sample) at a po-sition and irradiating with the laser beam. In inorganic samples, the Raman spectrum was found to be more distinct with reduced particle size. However when samples were provided as dispersed particles of sizes below the wavelength of the laser beam in a matrix (such as fillers in polymer matrix and droplets in emulsion) the reduction in Raman spectral signal was observed, such as in the case of titanium dioxide in poly-thene matrix.

The powdery samples generally are smaller than the laser beam itself and get burnt or illuminate causing reduction in Raman signals. This problem was encoun-tered by preparing the sample as a halide disk and placing it on the laser beam ex-cited with a power of about 1400 MW. The method was effective and it exhibited strong Raman signals with no burning of the particles. For solids which strongly burn on exposure to high-power laser beam, a spinning mechanism of the sample was de-veloped to eliminate the burning effect at a speed not greater than 50 Hz to prevent beat formation on the scattering output. However, in case of the NIR FT Raman spec-troscope, the spinning must be maintained at a speed less than 60 Hz.

To deal with the issue of fluorescence, the raw sample can be left exposed during the initial step for the fluorescence to get burnt out, which might take a few seconds to complete or even hours. With higher wavelength beam, fluorescence is observed to get reduced or even eliminated unlike the copper phthalocyanine where the fluorescence increases. This is why in FT Raman spectroscopy where the laser beam has an excitation wavelength of 1064 nm, fluorescence effect is of minimum concern. It must be kept in mind that the color of the sample has no significance in determining the fluorescence effect. Again it was observed that sensitivity of the Raman spectrum enhances with lower wavelengths up to the UV region though the sample has the tendency to thermally degrade. In case of liquids, the fluorescence effect is clearly dominant though burning effect is not present due to mobility of the liquid and transfer of energy to other molecules, thus enhancing the heat capacity.

Polymers of all types and forms generally exhibit a weak Raman spectrum signal, due to which they are usually employed as a container for other samples. However, a small amount of azo dye into the polymer film can exhibit Raman lines for both the dye and the polymer. Thin film polymers are generally recommended to be folded multiple times to eliminate any orientation effect but the thin film polymers that are too small to be folded can be placed across the mirrored face holder for strong Raman lines.

7.7 Data handling and interpretation

The handling of detected data and their interpretation are important aspects of Raman spectroscopy both for qualitative and quantitative analysis. A complete knowledge of the manipulation during the production of spectrum and its display techniques are essential for better and appropriate interpretation of the characteristics of the Raman spectrum. Instrumental features such as beam splitter and filters and their manipulations can contribute tremendously to the output characteristics of the Raman spectrum. In FT Raman spectrometers, the raw output signals are in the form of an interferogram, which is later computer-manipulated to present as a spectrum. Experienced spectroscopists need to look for the noise in the signal away from the main signal to know the relative intensities at the peak. Comparison between the intensities of different spectra by overlapping with each other is an important technique used to interpret the strength, quantity, and composition of the sample material. Expertise on spectrum manipulation is essential where overuse of the smoothening programs might land one into different conclusions or interpretations that are far from the real solution.

In terms of displaying the spectrum, the wave number is generally taken from low to high, whereas in most other cases as in the case of comparison with infrared spectrum, the wave number is taken from high to low so that both the Raman and infrared spectra can be overlaid and band positions compared. Spectral presentation generally has been seen to omit the effect of anti-Stokes lines and only the Stokes lines are studied.

Qualitative analysis represents the identification of the composition of materials or the study of their actual spectrum, while on the other hand quantitative analyses deals in the quantity or the amount of constituents present in a sample. Quantitative analyses therefore are concerned with the intensity of the spectrum. Various scales and tools that vary with different instruments and do not provide a quantitative measure of the spectrum but are commonly used for comparative measurement and band ratio quantitation are available. These are the spectrum scales and data handling packages for spectral enhancements [44].

7.7.1 Quantitative interpretation

The quantitative analysis of the Raman spectroscopy output signals generally are dependent on the conditions at which the experiment has been performed (the mirror angles, laser power, detectors, sample concentrations and volumes, cell window material, etc.). The same conditions are not possible for different experimental setups due to which calibration is an important step to be considered. Checks are being carried out in calibration with standard samples and their relative intensities at certain wave numbers for a single wavelength laser source such as barium sulfate having a strong band at 988 cm^{-1}, silicon at 520 cm^{-1}, and diamond at 1364 cm^{-1}. Relative intensities can be enhanced by resonance but are generally affected by self-absorption and temperature of the laser source. The refractive mediums between the sample and the laser source (light passing through these mediums), misalignment of the cells in the sample, and the depth of focus of the laser source also contribute towards changes exhibited in signal intensities. In order to combat these problems, a stable holder for the sample position relative to the collection optics is desired and instrumental parameters are set to be the same for all experiments. A perfect quantification of the scattered data can be obtained if all the above mentioned requirements are fulfilled. Moreover, quantification of the signal becomes easier with the use of a double beam approach.

For quantification, the most commonly used method for analyzing the intensities is to measure the peaks though the area under the peak can also provide accurate measurements in many cases. A baseline needs to be constructed in regions where there is no Raman spectrum in order to take into account the band shape and the neighboring bands. Though position of the base point is not critical, in calibration and test samples, the method for finding the point must be consistent. A number of software packages are available for quantitative analysis of Raman spectrum some of which are used for composition measurements following the methods of least square fit, principle component regression, and partial least squares. Packages for band resolution and spectral enhancements are also available in many instruments. Thus, software packages must be used with complete understanding of the requirements and their applicability for a particular problem to be studied.

7.7.2 Qualitative interpretation

Complete history of the sample nature, instruments employed, and experimentation technique is required before actually analyzing and interpreting the sample spectrum. Various factors involved such as the build of the molecules in the sample, laser wavelength and power, and accessories used for sampling play a crucial role in affecting the outcomes of the Raman spectrum. If proper steps are being followed and simple to complex factors kept in mind, interpretations can be made out of the spectrum, yielding physical, chemical, and even electronic information of the sample material. From a minute change in band position and intensity, huge information can be drawn out of the sample spectrum with correct interpretation. Interpretation of vibrational spectroscopy can facilitate detection of the composition of unknown materials, characterizing reaction by products and reaction followed. Raman spectrum as compared to infrared spectroscopy though simpler and clear, is tough to fingerprint due to its weak bands. Though in both IR and Raman spectroscopy, expertise and in depth knowledge in all the factors that influences the spectrum are required, Raman spectroscopy experiments involve comparatively lesser number of factors and complexities with minimum effort even in the preparation of samples. Instrumental effects can show up in the scattered spectrum such as the effect of cosmic rays, room lights, cathode ray tubes, and strip light that can be strong enough for weak spectra. One more important factor that needs to be considered is the nature of the spectrum of the particles in a matrix of different material as can be observed in a sample consisting of sulfur molecules in polymer matrix. Sulphur usually exhibits strong spectra while polymers exhibit the opposite, due to which sulfur spectral lines dominate the output signal with all its peaks. This phenomenon should not be interpreted for quantitative analysis as quantity of the sulfur has got nothing to do with the intensity. The physical state of the samples have an effect on the quality of the spectrum as can be seen that crystalline solids exhibits a sharper and stronger spectrum while liquid sample exhibits a weaker spectrum. A few other factors that can influence the quality of the scattered spectrum include temperature, pressure, orientation, crystal size, polymorphism, pH changes, burning, and fluorescence.

Thus sequential steps in knowledge of the sample, sample preparation effect, effect of the instrument/software, spectrum interpretation, and computer- aided spectrum interpretation are important for any spectroscopist before any experiments on Raman spectrum is conducted.

7.8 Examples of Raman spectroscopy

7.8.1 Raman spectroscopy on polystyrene nanocomposites with single-walled carbon nanotubes

The interaction between single-walled carbon nanotubes (SWNCTs) and polystyrene within the fabricated nanocomposites has been investigated with Raman spectroscopy. A shift to higher wavenumbers was observed for the G⁻, G⁺, and G' SWNCT bands with the incorporation of SWNCT into the nanocomposites [44]. The shift could be attributed to the transfer of the mechanical compression from polystyrene to SWNCTs. The transfer in mechanical compression was around 518 MPa.

The Raman spectra of SWNCTs, polystyrene, and nanocomposites of SWNCTs with polystyrene have been depicted in Fig. 7.6. It can be observed from the depiction that at 178, 1566, 1590, 1333, and 2667 cm^{-1}, the RBM band, bands for G⁺ and G⁻, and band for G' have been observed, respectively. The revelation that SWNCTs are semiconductors can be made from the shape of the G bands. The Raman spectra associated with the SWNCT and polystyrene nanocomposites dominate over the Raman spectra of SWNCTs and polystyrene in its pristine state. Moreover, it can also be observed that the intensities associated with the Raman bands for pristine polystyrene is weaker. The appearance of Raman peaks for SWNCTs in its nanocomposites with polystyrene is at higher wave numbers in comparison to the Raman peaks for the SWNCTs in its pristine state. For instance, the shift of G bands from 1566 and 1590 to 1568 and 1593 cm^{-1} can be observed from the depiction in Fig. 7.6.

An investigation into SWNCTs nanocomposites with polystyrene for its different loadings, that is, 0.5, 1, and 3 wt% was also made using the Raman spectroscopy. Raman spectra have been depicted in Fig. 7.7 (a–c). The three regions, that is, 275–50, 1700–1450, and 3000–2500 cm^{-1} have been observed for the SWNCTs in their pristine state and for the three different kinds of SWNCTs nanocomposites. Two peaks at 178 and 164 cm^{-1} can be observed in Fig. 7.7 (a) for the RBM band of SWNCTs in its pristine state. A shift of the stronger peak at 178 cm^{-1} can be observed for spectra associated with the 0.5, 1, and 3 wt% SWNCTs nanocomposites. The shift took place from 178 cm^{-1} to 187, 184, and 182 cm^{-1}, respectively, for the three different kinds of composites. A solid dark circle in Fig. 7.7 (a) points to the weaker one situated at 164 cm^{-1}. A shift to higher wavenumbers can be observed for the weaker 164 cm^{-1} for the polystyrene nanocomposites with 0.5 and 1 wt% SWNCTs. However, this can be seen to be missing for the Raman spectra for 3 wt% SWNCTs polystyrene nanocomposites. This could be attributed to the presence of SWNCTs with nonuniform diameters.

A comparison of the G-band regions for the pristine SWNCTs and their nanocomposites with polystyrene has been made in Fig. 7.7 (b). Two sub-bands of G⁻ and G⁺ forms the G band for the SWNCTs. These sub-bands arise because of the vibrations of carbon along the circumferential and axial directions of SWNCTs. Raman spectra corresponding to the G' band region can be observed in Fig. 7.7 (c). The stretching modes

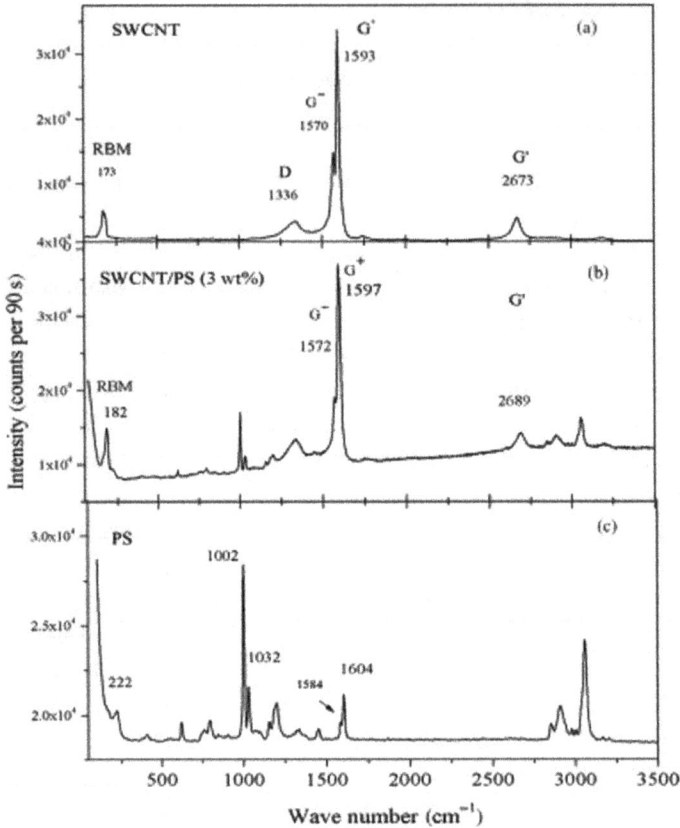

Fig. 7.6: Raman spectra of (a) SWCNTs, (b) 3 wt% SWCNT/PS, and (c) PS (reproduced with permission from [45]).

associated with the symmetric and antisymmetric ν_s (CH$_2$) could be observed for the peaks at 2850 and 2904 cm^{-1} and have been reflected in Fig. 7.7 (c) with a solid dark square. A shift to higher wave numbers for the G and G' bands can been observed from Figs. 7.8 and 7.9. The shift to higher wave numbers has been observed to be similar to shift behavior corresponding to RBM. It could also be observed that the SWNCTs nanocomposites with 0.5 wt% result in larger shifts in comparison to the nanocomposites having 1 and 3 wt% SWCNTs.

The following reasons may be attributed to the shift in the Raman bands associated with SWNCTs: chemical interaction arising between polystyrene and SWNCTs, changes arising because of the tube-tube interactions owing to the SWNCTs distribution, and mechanical compression from the polystyrene matrix. The first reason may arise due to the doping phenomenon that may promote the charge transfer or due to the formation of the chemical bonds between the carbon nanotubes and the polymeric polystyrene matrix. These two factors may result in the shift to higher wave numbers of the Raman peaks for the carbon nanotubes by several tens of cm^{-1}, or there may arise a shift to lower wave number by the similar magnitude. CH interactions between the carbon nanotubes with π-electron-rich surface may be the other possibility.

IR spectroscopy can be employed in order to study such type of interactions. This is because the shifting of the CH_2 band takes roughly by 10 cm^{-1}. IR spectra of the pristine polystyrene reveal the location of CH_2 bands at 2924, 2850, and 1371 cm^{-1} [44]. However, no shift in the CH_2 has been revealed, which indicates the absence of CH interactions and hence does not play any vital role in such kind of samples. This signifies that the observed Raman shifts cannot be explained with the charge transfer phenomenon. Moreover, it has been also revealed that the van der Waals interactions that occur amongst the bundles of SWNCTs are also not suitable to provide justifications to the observed Raman shifts within the system.

Therefore, the study concentrated on the vibrations arising due to elongation and shrinkage of carbon atoms. The vibrations result in Raman shift of SWNCTs by several wave numbers. Hydrostatic pressure is experienced by the SWNCTs from the liquid or polymer in which these are embedded. The Raman shift can be attributed to this hydrostatic pressure. An investigation into the effect of different magnitudes of high pressure on the SWNCTs has been investigated comprehensively by Loa [45] using Raman spectroscopy. It has been revealed through the investigation that a reversible change within the structure of SWNCTs was created through the application of pressure. Therefore, due to the high pressure, the SWNCTs could be transformed to diamond and other super hard phases. In the study conducted by Yan et al. [44], it has been revealed that SWNCTs underwent mechanical compression from the surrounding polymeric matrix and the magnitude of the same was estimated to be 518 MPa. This has been revealed from the Raman shifts of SWNCTs and SWNCTs/polystyrene nanocomposites (1 wt%).

The Raman shifts for the G' and G bands at different positions within the nanocomposites have been depicted in Figs. 7.8 and 7.9, respectively. Relatively more shift in G^+ bands has been revealed in comparison to the G^- bands. This observation reveals that the mechanical pressure had a more dominating effect on the carbon-carbon atomic vibrations along the axial directions in comparison to the circumferential vibrations. Therefore, it can be deduced that the one of the excellent ways to analyze the SWNCTs interactions with polystyrene is the observation of the shift behavior associated with G'. Strain-dependent spectroscopy has been employed in one of the studies to detect the stress transfer from polymeric matric to CNTs. The G' band experiences a large downshift when the CNTs undergo strain. Larger shifts have been revealed in case of lower SWNCTs loading within the polymeric matrix. This is because of the fact that lower loading levels result in better alignment of the SWNCTs. Aggregation occurs for higher loading levels of SWNCTs in the polymeric matrix. As a result of the aggregation phenomenon, lower mechanical compression is experienced by the SWNCTs in comparison to the uniformly distributed SWNCTs.

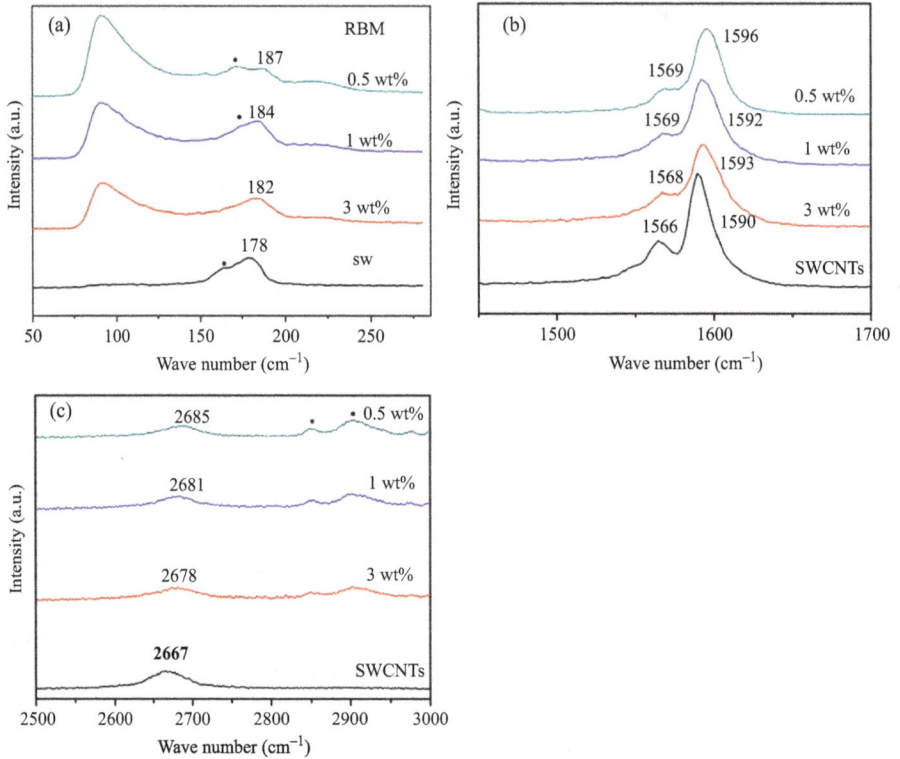

Fig. 7.7: (a) RBMs, (b) the G Band, and (c) the G' band region in the Raman spectra of SWCNT/polystyrene nanocomposites (reproduced with permission from [45]).

7.8.2 Raman spectroscopy on analyzing the interfacial load transfer in case of graphene-based nanocomposites

An investigation into the interaction between graphene platelets and Polydimethyl-siloxane (PDMS) matrix has been approached using the strain-sensitive Raman spectroscopy in one of the studies conducted by Srinivas et al. [46]. Large debonding strains amounting to approximately 7% for graphene platelets in PDMS have been revealed. The results in the study indicated enhanced load transfer from the matrix to graphene platelets. Raman spectroscopy was adopted by Srinivas et al. [46] to investigate the load transfer in PDMS matrix encompassing graphene platelets and SWNCTs. The investigation was made under compressive as well as tensile loading. Improved load transfer for graphene platelets has been revealed in comparison to that obtained for the SWNCTs.

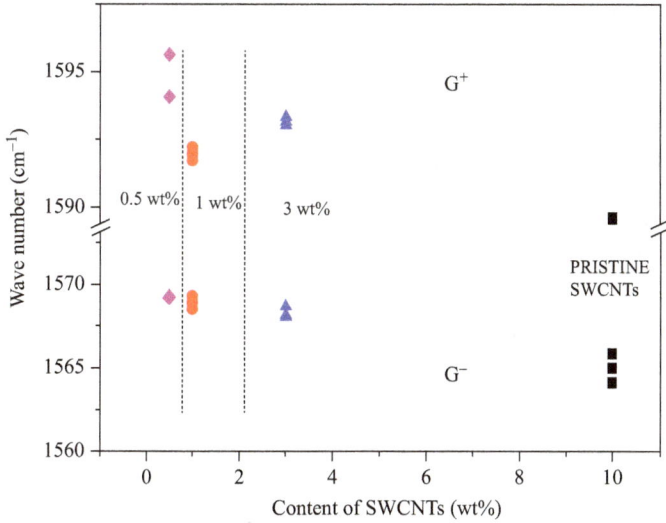

Fig. 7.8: Raman shift of the G⁺ and G⁻ bands for SWCNT/PS composites (reproduced with permission from [45]).

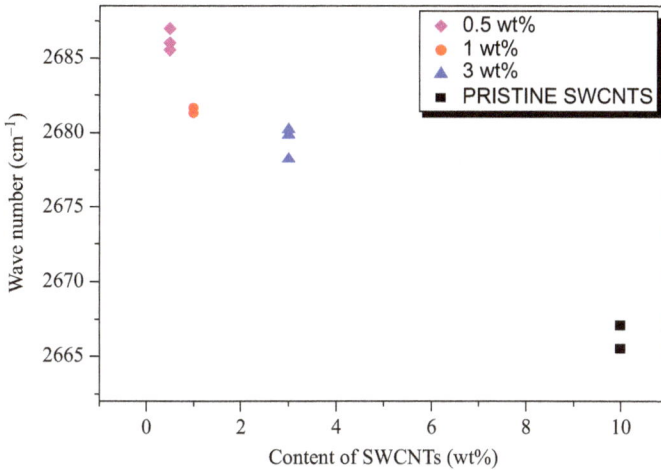

Fig. 7.9: Raman shift of the G' bands versus the SWCNT loading in their nanocomposites (reproduced with permission from [45]).

Figure 7.10 depicts the morphology associated with the graphene platelets-based PDMS nanocomposites. The graphene platelets can be observed as wavy edge structure. It is observed that for strain levels close to 7% (Fig. 7.11) the graphene platelets debonds from the PDMS matrix. The high strain to debonding reveals a stronger interface between the graphene platelets and the PDMS matrix.

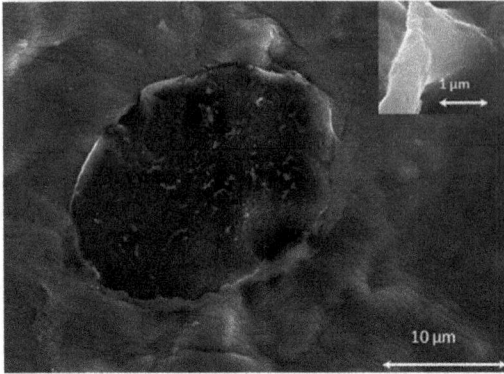

Fig. 7.10: SEM image of graphene platelet in PDMS matrix (reproduced with permission from [46]).

The Raman shift of the G-band peak has been revealed in the Raman spectroscopy of graphene platelets-based PDMS nanocomposites with the application of the strain (Fig. 7.12). A shift in the G-band of the graphene platelets have been revealed to be a function of the strain applied. It has been observed that the peak reverted back to its original peak position once the debonding of the graphene platelets took place with the PDMS matrix.

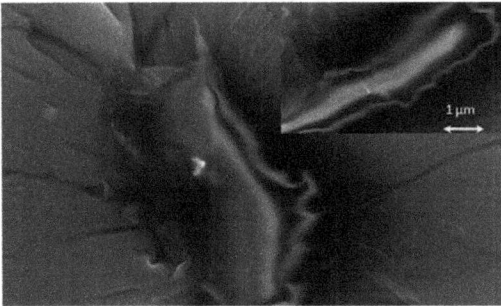

Fig. 7.11: Interfacial debonding at a strain level of ~7% (reproduced with permission from [46]).

A comparative analysis for the G-band peak shift for the graphene/PDMS, graphene/polystyrene, and SWNT/PDMS composites has been depicted in Fig. 7.13.

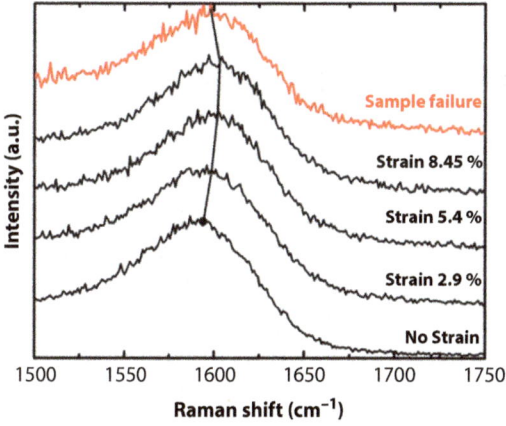

Fig. 7.12: Response of G-band shift of graphene platelets-based PDMS nanocomposites to applied strain (reproduced with permission from [46]).

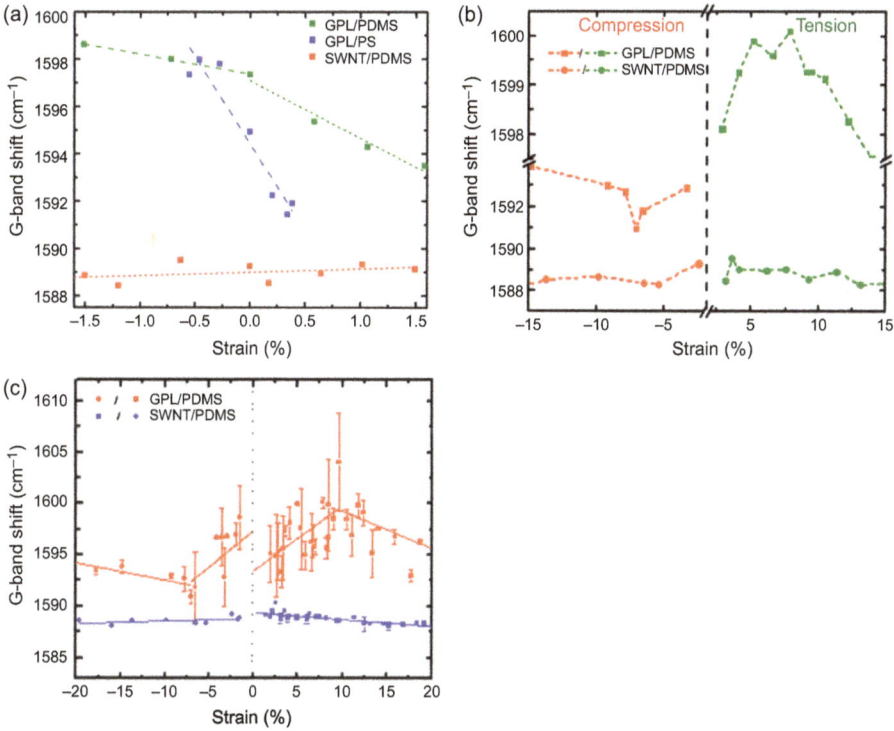

Fig. 7.13: (a) Comparison of G-band shift for nanocomposites with below 2% applied strain, (b) greater than 2%, and (c) best fit plot (reproduced with permission from [46]).

7.9 Conclusions

The present chapter provided an overview of Raman spectroscopy. Instrument detailing for generating Raman spectra and sample preparation aspects have been covered after a short introduction section on the Raman spectrometer. Handling of sample is also one of the quintessential aspects in Raman spectroscopy and the same can aid the technicians, instructors, and researchers in handling the sophisticated instrument with due care. The data interpretation has then been detailed next, covering quantitative as well as the qualitative techniques. The chapter then discusses a few relevant case studies to better comprehend the data generated using Raman spectroscopy.

Chapter 8
X-ray photoelectron spectroscopy

8.1 Introduction

Heinrich Hertz first observed around the year 1887 that when an electrically insulated metallic object is irradiated with a light source under vacuum condition, there is generation of a spark on its surface. This effect was termed as the "Hertz effect," which was later explained by Albert Einstein in 1905 as the effect observed when the energy of the photons (energy packets with zero mass) striking the metallic surface exceeds the binding energy of the electrons to its respective nucleus and termed as the photoelectric effect for which he received the Nobel prize in Physics in 1921. Photoelectrons emitted are accompanied by emission of photons (fluorescence) or Auger electron emission. The kinetic energy of the released photoelectron is measured, which is a function of the binding energy of the electron to its core and is discrete in nature.

Kai Siegbahn and his coworkers first developed an XPS instrument to analyze the core photoelectron emissions with a sufficient high resolution for carrying out speciation analysis in 1967, for which he too received a Nobel Prize in Physics in 1981. The development of the XPS was late due to unavailability of any equipment that was able to create high vacuum conditions required for X-ray transmission in the analysis of the photoelectron emission from a sample surface. With further research, developments and understanding of the process observed in XPS instrumentation has led to the identification and quantification of the elements present in any material (solid or liquid), enhancement in the sensitivity to material concentrations, high resolution in surface speciation, and ease of analysis. In this chapter, XPS instrumental details and working, XPS sample preparation, data analysis, and interpretation have been discussed in detail.

8.2 Instrumental details and working

The instrumentation of an X-ray photoelectron spectroscopy (XPS) [47, 48] involves a few components, namely, the vacuum generator, X-ray sources, extraction optics, and a detection system. In order to make the best out of these instruments the spectroscopist needs the skill and knowledge of conducting a successful analysis of the spectra produced, collecting data and interpreting the best outcome possibility. As this process involves photoelectric effect involving transfer of electrons from the cathode to the anode, the inside environment of an X-ray spectroscope must be free from any air molecules, electric, or magnetic field that can interact with the kinetic energy of the moving electron. However, these fields do not have any effect on the X-ray beam as it has no charge. To provide shielding from these extraneous magnetic and electrostatic

https://doi.org/10.1515/9783110997590-008

fields, Mu-metals, which include 76% Ni, 17% Fe, 5% Cu, and 2% Cr that are generally used for the construction of energy filter housing, analysis chamber, detector housing, and the lining of these chambers/housing.

8.2.1 Vacuum systems

Vacuum conditions are required in an XPS system since at ambient temperature and under atmospheric pressure there are 2×10^{19} molecules/cm^3, which can prohibit the transmission of the electrons from the sample to the detector.

It was not possible to create effective vacuum conditions better than 10^{-7} Torr though the first vacuum pump was discovered in 1650. It was only around the 1950s that the vacuum conditions were improved and an effective XPS was developed in 1960.

Fig. 8.1: UHV XPS system.
Source: UHV Multichamber XPS (2012), Prevac (reproduced with permission from [49]).

Vacuum is basically produced by removing the air molecules from a confined volume of it or reduction in pressure (which is basically collision of the air molecules onto walls of the confined volume). For an ultrahigh vacuum (UHV) condition in an XPS (Fig. 8.1), a highly specialized pumping system that can produce a least pressure of below 10^{-1} Torr is required. Vacuum chambers in an XPS are mainly constructed out of stainless steel with linings of Mu-metal or of the Mu-metal itself which has a number of advantages such as low outgassing rate, low vapor pressure, low corrosion rate, cost effectiveness, and ease of fabrication. The materials that have high vapor pressure (such as borosilicate glass, epoxy resins, rubber O ring, adhesive tapes) are avoided to eliminate adsorbent generation. For sustaining the UHV conditions, chambers are usually manufactured in the shape of a sphere or cylinder onto which ports are attached through which sample access is provided using specialized welding procedure. Other units are joined using flange/gasket arrangement. Copper gaskets are used, which are pressed between two flanges and must be replaced after one use. To avoid this, other soft gaskets such as Al, In, and Au are also generally used for UHV conditions. Carbon-based O rings are used – one that is reusable for HV applications. Samples are introduced into the chamber via an introduction chamber sealed by a series of valves between the atmosphere and the vacuum chamber. Pressure can be reduced to 10^{-7} Torr within 10–20 min and the valves are then opened for the introduction of the sample into the chamber. Vacuum gauges are installed that measure the vacuum produced in the chamber. Vacuum pumps can be divided into three groups which are: positive displacement pump, momentum transfer pumps, and entrapment pumps. Positive displacement pumps are generally rotary vane pumps that can reduce the pressure of the chambers from atmospheric pressure to about 10^{-3} Torr, which is followed by the momentum transfer pumps, further reducing the pressure to 10^{-10} Torr. The commonly used momentum transfers pumps are turbomolecular pumps, which are used with the positive displacement pumps and works in a condition that is already below 10^{-3} Torr. Entrapment pumps are used, which overcome this drawback of the momentum transfer pumps and are able to drive away air molecules, reducing the pressure from the atmospheric condition to 10^{-7} conditions in one go.

Pumping speed of the pumps are an important criteria while developing UHV conditions in the chambers, which is basically dependent on a number of factors. These factors are the chamber volume, outgassing rate, and the desired least pressure. Pump down times are generally dependent on the removal rate of the surface adsorbents from the metal surface of the chamber. These are therefore baked at a temperature of around 150–200 °C for a period of 12–48 h whenever it is contacted with atmospheric air. Pressure measuring gauges are used as a combination of Pirani gauges and ionization gauges of which the useful pressure range is between 10 and 10^{-3} Torr for Pirani gauge and between 10^{-3} and 10^{-10} Torr for ionization gauge. It is always made sure in the design of the chamber that all the abovementioned equipment do not come in the line of sight of the sample surface and the detectors.

8.2.2 X-ray sources

High energy photons in the wavelength order of the X-ray region of the spectrum are an important source for analyzing the surface characteristics of a sample. These photons can penetrate into the core electron level of the atoms on the surface to knock out electrons to be detected. X-rays in an XPS system are produced from the sources that are described below: X-ray tubes, which include monochromatic sources and standard sources based on their geometries and synchrotron sources. In X-ray tubes, the high energy electrons are made to accelerate from the cathode and directed onto a metallic plate, which is basically an X-ray anode. Cathodes generally are thermionic sources and anodes are commonly made of Al metals. The use of Al as an anode has a number of advantages such as relatively high intensity and energy of Al-Ka X-rays, minimal energy spread of Al-Ka X-rays, effective heat conductivity, and ease of manufacture and use. In thermionic sources, electrons are generated by temperature increase to that point of overcoming the binding energy of the electrons that gets passed on through vacuum and directed onto the anode, releasing X-rays in the process.

In order to maximize X-ray flux, high energy electrons must be impacted on the Al anode of which only a small percent of the deposition energy is used up in producing x-rays. The remaining energy in the production of secondary electrons or Auger electrons and generation of heat is consumed. It is due to this, sufficient conductivity is required of the anode metal to drive away the heat produced and in some cases water (deionized water to prevent electric discharge issues or arcing) cooling may also be required.

8.2.2.1 Standard sources

Standard sources generally include dual Al and Mg anodes, which are water-cooled and attract negative charges or electrons to impinge on its surface and emit unfocused beam of high energy X rays. A few advantages of standard sources over monochromatic sources include low cost and ease of switching between anode materials. Disadvantages include a larger spread of energy, presence of noisy peaks, sample degradation due to heat generation at the area of impact of the high energy electrons, and lower photon flux per unit area.

8.2.2.2 Monochromatic sources

Monochromatic sources (Fig. 8.2) include all the elements present in a standard source in addition to a concave single crystal, which focuses the beam of X-ray onto the surface of the sample. A commonly used material in the commercially available XPS system for concave single crystal is the quartz since it is relatively inert to atmosphere, exhibits better compatibility for UHV conditions, and has low degradation. The anode is water-cooled as in the case of the standard sources. The advantages of monochromatic source over the standard source are its focused beam with no requirement for the source to be placed near the target and a narrow spread of energy. The disadvantages when compared to the standard sources include its high cost, higher sample charging in insulating surfaces, and limited number of anodes allowed for satisfying Bragg's criteria.

Fig. 8.2: Monochromatic light source.
Source: Monochromatic Light source for Gems Refractometer (2020), optics factory
(reproduced with permission from [50]).

8.2.2.3 Synchrotron sources

Synchrotron sources generally employ the cathode ray principle to generate the electrons, which are accelerated in a circular path using the magnetic field, and the photons are emitted tangential to the electron beam. The X-ray is thus produced, which has the intensity that is dependent on the radius of curvature of the electron beam and is inversely proportional to the cube of the electron beam energy. Synchrotron sources have advantages over other X-ray sources in that the photon energy can be selected, high photon flux per unit area than in monochromatic or standard sources, collimated beam that can be focused to 0.1 micrometer, narrow spread of energy, and polarizable beam interacting with the sample. Disadvantages include high degradation rate of the sample especially the organic samples, unavailability, and high cost.

8.2.3 Energy analyzer

Effective speciation analyses of the information obtained in an XPS system are required to enhance the resolution and transmission for better separation between the closely spaced peaks and maximization of the sensitivity, respectively. Energy filter configurations that are used in XPS include Cylinder Mirror analyzer (CMA) and Concentric hemispherical analyzer (CHA).

8.2.4 Cylindrical mirror analyzer (CMA)

CMA (Fig. 8.3) generally consists of an arrangement of concentric cylinders one over the other between which potential difference can be developed. When bought near to the sample surface, electrons emanate out at an angle of 47.7^0 and higher transmission is obtained as compared to CHA, enhancing kinetic energy sensitivity to sample surface position and lowering energy/angle resolution. The energy resolution is dependent on the dimension of the slit present on the inner cylinder during the flight path of the electrons, which lies between 1–0.1 eV. The voltage applied between the cylinders can be given as:

$$V = \left(E_0 \left(\ln \frac{R_{out}}{R_{in}} \right) \right) K_{CMA} \tag{8.1}$$

where E_0 is the energy of the electrons, R_{in} and R_{out} are the radii of the inner and outer cylinders, respectively, and K_{CMA} is a geometric constant. In a single-pass CMA, the electrons intersect the cylinder axes once while in a double- pass CMA, electrons intersect the cylindrical axis twice.

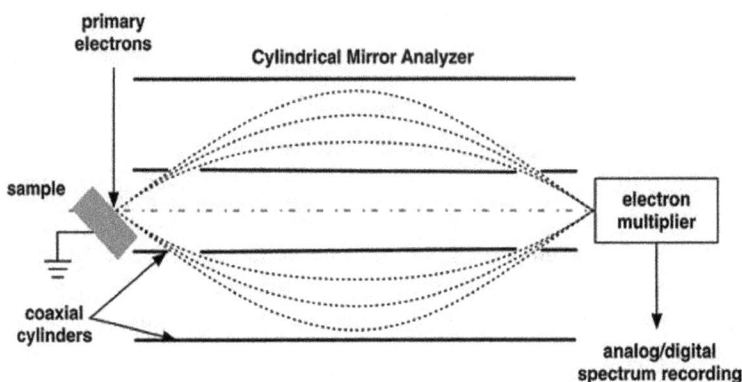

Fig. 8.3: Cylinder mirror analyzer.
Source: Auger electron spectroscopy (2013) (reproduced with permission from [51]).

8.2.5 Concentric hemispherical analyzer (CHA)

CHA (Fig. 8.4) generally contains two hemispheres of different radius (R_{in} and R_{out}) one over the other, through which electrons are passed. The path of the electron remains tangential to the average value of the outer and the inner radius of the hemispheres (R_{el}). Specific potentials applied to the hemispheres deflect the electrons of a specific energy (E_0) onto the detector, which can be given as:

$$V_{in} = E_0 (3 - 2(R_{el}/R_{in})) \tag{8.2}$$

$$V_{out} = E_0 (3 - 2(R_{el}/R_{out})) \tag{8.3}$$

where R_{el} is the average radius equal to $(R_{out} + R_{in})/2$.

Cylindrical mirror analyzer generally exhibits a higher energy resolution of the order of values better than 0.01 eV as compared to the CMA. The CHA exhibits poor

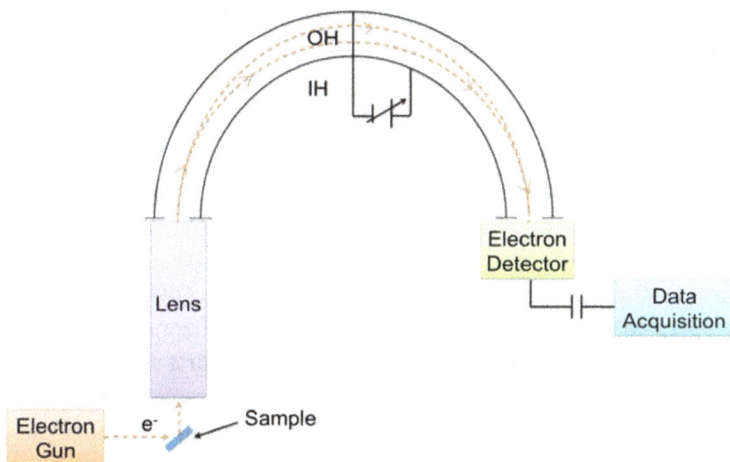

Fig. 8.4: Concentric hemispherical analyzer.
Source: Schematic of concentric hemispherical analyzer (reproduced with permission from.[52]).

influence of the distance between the sample and the analyzer on the kinetic energy with value of 0.1 eV/mm and therefore can analyze larger areas than CMA. However, CHA exhibits poorer transmission due to limited collection solid cone than a CMA, which is compensated to some extent by installing transfer lens prior to energy analyzer and using multichannel detectors. Transfer lens also displays a greater accessibility to the sample surface. Energy spectrum in XPS is obtained via two modes, which are by scanning the potential of the inner and outer hemispheres across E_0 range also known as a constant retard ratio (CRR) or the fixed retard ratio (FRR) mode, and accelerating and decelerating the electrons to specific energy (E_0) before passing through the CHA/second stage of a double- pass CMA also referred to as constant analyzer energy or fixed analyzer transmission. The former technique is applied for single-pass CMAs while the latter one is applied for CHA and double-pass CMAs. Acceleration and deceleration of the electrons is achieved by a transfer lens.

8.2.6 Detectors

After measuring the energy of the electrons emitted through the energy analyzer, the number of electrons emitted, which have an intensity effect for interpretation by detectors needs to be measured. These two parameters are used to construct an energy-intensity graph. It is very important for a detector to be sensitive and it should be able measure individual electrons that are emitted from the sample in pulse. It is measured in units of current and represented as counts per second. Since in a conventional pulse counting device the sensitivity generally lies in the 10^{-15}A range, which is equivalent to charge carried by 6241 electrons per second, it is very important to multiply the number of electrons or a gain in signal for higher sensitivity to individual electrons.

8.2.7 Electron multipliers

Electron multipliers (EM) are thus generally used to enhance the conventional sensitivity and can be operated at UHV conditions required by XPS. EM amplifies the electron count (or the signal) by allowing electrons to impinge on its first surface or part and emitting many other electrons for higher sensitivity of a pulse counting device. The different configurations of an EM are given as discrete dynodes, channeltrons, and Micro channel plate (MCP) detector. The pulse generated can then be collected via an anode collector, a delay line detector, or a phosphor screen.

Discrete dynodes EM (Fig. 8.5) generally consist of conversion electrode (dynode) spanning an area of 10×10 mm^2 followed by numerous acceleration dynodes. The electrons first strike the conversion dynode producing other secondary electrons, which are then accelerated to the next diode and then the next. The number of electrons exiting out of the last dynode thus is a multiplication factor of the original electrons. Though the gain in a discrete dynode configuration is up to 10^9, the dynode detector tends to get saturated for an incoming electron count rate exceeding 10^6 cps resulting in a loss of linearity in the input-output signal.

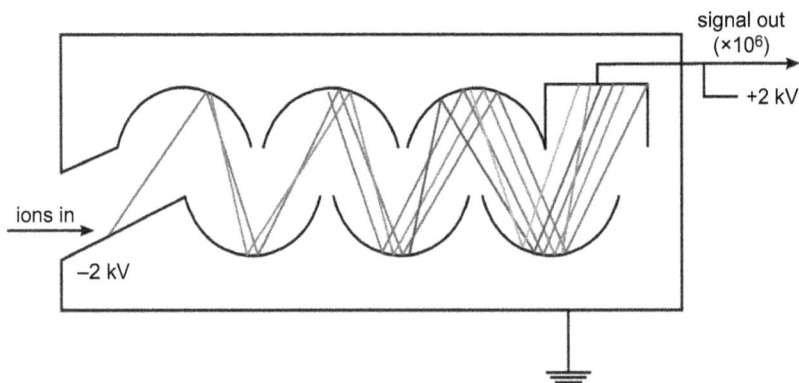

Fig. 8.5: Discrete dynode EM.
Source: discrete dynode secondary electron multiplier (2016) (reproduced with permission from [53]).

8.2.8 Channeltrons

Channeltrons (Fig. 8.6) are continuous dynode segments that form a horn shape having two perpendicular openings through which the electrons can be directed. It has a similar working operation as in the case of a discrete dynode EM with an electron-emissive material such as PbO coated on its inside surface. The electrons are directed into the horn-shaped structure by applying a potential difference of about 2–4 kV across both ends.

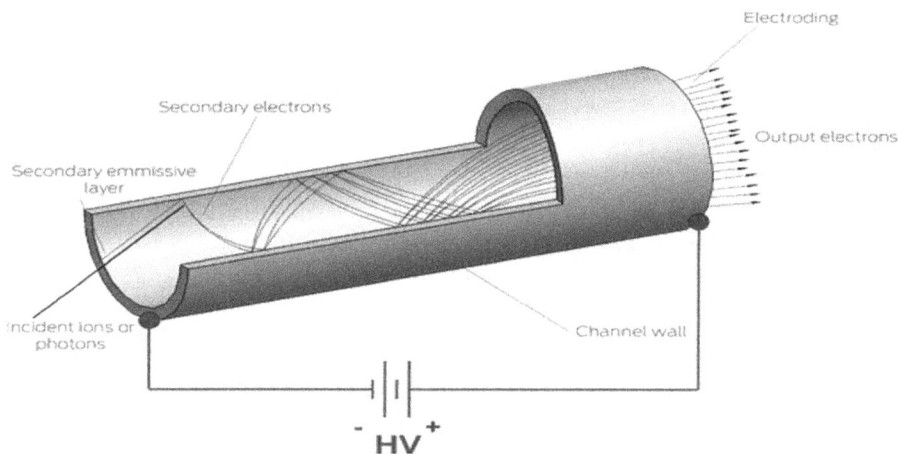

Fig. 8.6: Channeltrons.
Source: Channeltron (retrieved on 25 February 2020), Photonis (reproduced with permission from [54]).

MCP detectors (Fig. 8.7) are parallel glass channels, which can be taken as an extension of one-dimensional channeltron arrays with inner diameter ranging from 10 to 25 micrometers and axis lying at an angle of 7^0 with the path of incoming electrons. The inner surface of the channels are coated with an electron-emissive material as can be seen in the case of channeltrons onto which incoming electrons strike to form cascading of numerous electrons to exhibit a gain of 10^4. Use of two channels back-to-back with one rotated 180° to the other can further enhance the gain to about 10^6; these are also referred to as chevron MCPs and have a saturation incoming electron count of 3×10^6 cps.

8.3 Sample preparation

Sample preparation in an XPS instrument is not tedious and is less time- consuming with few requirements such as the size of the sample that must be suitably fitted to the sample platform, the surface of the sample must be clean, and that the surface of the sample must be compatible with X-rays and UHV conditions.

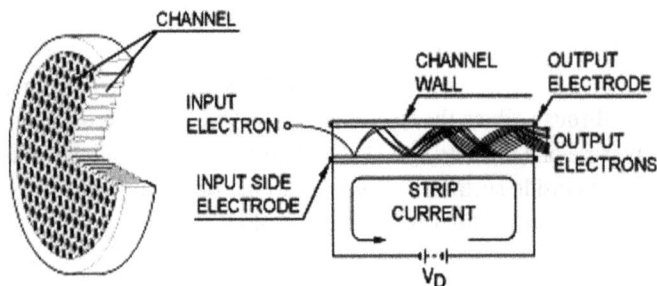

Fig. 8.7: Micro channel plate.
Source: Schematic representation of the operating principle of a microchannel plate (MCP)
detector (2018), University of Groningen (reproduced with permission from [55]).

The sample size and geometry should optimally be that of a small coin, which facilitates ease of handling and allows acceptable pump down speed due to its small surface area. However, greater pump down speeds, larger introduction chamber, and larger analysis chamber are required in the case of a larger sample size. If the samples are larger than a few millimeters then it can be simply clipped onto the sample platform, while for sample sizes smaller than the few millimeters or in powdery form can be formed into coin-shaped and similar geometry. The sample powder is generally put on an indium strip and compressed using another indium strip or using a spatula, which is then clipped onto the sample platform. Another method of sample presentation is to sprinkle the powders over a double- sided spectrum-grade sticky tape placed on a flat disk, which is then mounted onto the sample platform. Another way of presenting nano-powder samples is to disperse them into an inert solvent and pipet onto a substrate and dry them. It must always be made sure that no loose particles exist, which can be ensured by inverting the platform or by using pressurized N_2.

Cleanliness is an important factor while working on a sample, prior to or during handling due to its influence on the signal of interest. Appropriate gloves preferably made of polyethylene and latex must be worn, though some latex gloves are not suitable due to presence of silicones that can segregate on the surface of the sample. Tools such as tweezers, spatulas, screw drivers, and sample support holders including the ones for mounting the samples and measuring the size of the sample must also be kept clean even before use. Cleaning of the tools is mainly accomplished using acetone, isopropyl alcohol, methanol, or by employing ultrasonic cleaning.

After mounting the sample onto the sample platform, the arrangement is introduced into the introduction chamber, which is a small chamber that facilitates quick pump down of air molecules in a time period of about 10–15 min for a clean and nonporous sample. For a porous material and large samples, the pump down time exceeds this limit and can even be continued overnight with UHV conditions reaching a pressure below 10^{-7} Torr. When the required conditions of the sample surface and pressure are reached, it finally enters the analysis chamber. UHV conditions help in re-adsorption of the gas phase contaminants onto the surface area of the sample that is to be examined under X-rays, electron, and ion impacts. In case of high vapor pressure sample however, a cooling system is necessary to be used to prevent sample outgassing, which tends to scale up with temperature. Cooling is basically done by externally located LN_2 Dewar being thermally connected to the sample platform within the instrument in UHV condition. High cooling also requires the use of liquid helium.

Cleaning of the sample surfaces in some cases can also be carried out by rinsing it in specific solvents and reagents before placing into the introduction chamber or heating the sample at low level of oxygen in the analysis chamber to eliminate carbonaceous layers from thermally stable materials. An alternative method of cleaning a specific area of the sample in the analysis chamber has also been performed using sputtering with low energy ions like Ar^+ ions. Cluster beam sputtering has also been widely accepted to remove organic layers and substrate from the sample surface with damage reduction.

8.4 Data analysis and interpretation

Before going for analysis of the sample surface using XPS, a few factors that need to be taken care of and measures that need to be taken for a high-performing analysis are: pre-alignment procedures, energy calibration, charge compensation, X-ray/electron-induced damage, and background signal.

8.4.1 Pre-alignment

Pre-alignment generally considers the optimal positioning of the surface region of interest on the sample and whether the surface region is in the field of view. *Charge compensation* is required for XPS analysis of an insulating material due to loss of electrons and buildup of positive charge on the surface of the irradiated sample, which affects the binding energy and kinetic energy of the subsequently released electrons. Electron flood guns are generally used to tackle this problem by shooting low-energy electron beam, which is a self-compensating system as low-energy electrons are deflected by the surface electrons to the affected zones once the target surface is charged. *X-ray/electron -induced damage* to the sample is evident in the XPS instrument, which arises from the energetic collision of the X-rays onto the target surface. Some of the effects on the sample commonly encountered due to X-ray irradiation are reduction of atoms/ions within the area analyzed, phase changes at the target area, desorption of surface species, and migration of the specific elements on the sample surface. *Energy calibration* of an XPS instrument is generally performed using narrow core level Cu 2p3/2, Ag-3d5/2, and Au-4f7/2 peaks from the sample emission under a high-energy resolution condition. Surface oxides must be removed before calibration to ensure potential errors in the enhancement of signal intensities. C-1s peak arising out of adventitious carbon helps in accurate calibration of the B.E.$_{xps}$ in the analysis of insulators.

8.4.2 Background signals

Background signals are the most evident in the XPS analysis of transition metals due to inelastic scattering of electrons, which results in nondiscrete energy loss. In order to remove the background from the concerned peak, the three commonly used subtraction routines are the linear routine, Shirley routine, and Tougaard routine.

8.4.3 Quantification

Quantification of the recorded signal in an XPS instrument can be carried out without much difficulty and without reference materials. Factors that can deteriorate the accuracy (to not better than 5–10%) and increase of errors to about 30% are the unaccounted background, concentration variation in the sample region of interest, and inaccurate electron Inelastic Mean Free Path (IMFP) values. The inhomogeneity in the concentration of the sample surface is mainly due to microscopic inclusions, macro region having different concentration, and thin film formation of different composition that is lesser than the sampling depth. By assuming a homogeneous layer of amorphous mixture during quantification of the XPS sample over the region of interest and employing a reference material having the same composition as that of the sample, the abovementioned factors can be solved. Quantification can also be performed without the use of reference material by considering the equation for intensity (I) of a particular photoelectron peak in a nonelastic scattering event as follows:

$$I = J c_a \alpha_{pc} K_f \lambda_{IMFP} \tag{8.4}$$

where λ_{IMFP} is the IMFP of the photoelectron, K_f represents all the factors in the instrument (transmission function), α_{pc} is the cross section of the photoelectron, c_a is the photoelectron concentration that initiates building up of ions/atoms on sample region of interest, and J is the x-ray flux striking the concerned sample region of interest. The parameters α_{pc}, K_f, and λ_{IMFP} are generally transparent to the analyst that can be considered appropriately in the instrument software while the parameter J remains constant throughout the analysis. "I" is generally taken as the bounded area by the signal peak and the background combined.

The photoelectron in the detector displays a graph between the intensities and the binding energy (B.E.$_{xps}$) on the detector. B.E.$_{xps}$ basically represents the number of protons in the nucleus and the distance of the electrons from the nucleus with a dependency of $(1/r^2)$ where "r" is the orbital distance of the electrons. Alteration and shifts in the values of the B.E.$_{XPS}$ are basically due to the free ions/atoms produced during colli-sion of the X-rays with the sample surface and binding on to the other elements in the sample. The cause of these shifts can be interpreted as the condition of the atoms that remains bounded in the sample revealing the speciation of the elements present. The cause for the change in the B.E.$_{XPS}$ of a particular element in different chemical environ-ments can be described as arising from initial state effects and final state effects.

Initial state effects generally comprise the bonding of the ions/atoms with other elements/atoms on the sample, which induces an increase in density of all electrons (including valence and core electrons) described by the charge potential model. These effects enhance the ability of an XPS instrument in determining the speciation of the photoelectrons emitting atoms/ions. The difference in B.E.$_{XPS}$ can thus be related firstly to the electronegativity (EN) of the neighboring atoms/ions, which applies only for sys-tems displaying similar chemistries and secondly to the bond distance or atomic radii, providing direct measure of the electron density, though hard to obtain. Initial state effects can be further subdivided as those arising from ground-state polarization that includes interatomic effects and intra atomic effects and spin orbit splitting.

Final state effects comprise the deviation of the electronic structure caused as a result of photoelectron emission when the core level electrons are involved. These ef-fects as are also influenced by the initial electronic structure or bonding with other ions/atoms nearby can significantly contribute to the identification of the speciation of the photoelectron-emitting atoms/ions. Final state effects can be subdivided into those arising from excited-state polarization effect that includes coulombic effects (intra-atomic) and multiplet splitting effects (intra atomic) and rearrangement effects. Rearrangement effects include the excitation of the valence electrons induced by the core hole, which is eventually followed by rearranging themselves into some relaxa-tion different state. Thus new features can be introduced in the XPS spectrum, such as shake up satellites, plasmon features, Auger peaks, shake-up satellites, and peak asymmetry. Final state effects are always active due to their complicated variations that also is a function of the core hole lifetime and the Auger process decay, which is related to the irreducible width of the photoelectron peak.

8.5 Examples on XPS

8.5.1 XPS study of PPy

The formation of Polypyrrole (PPy) was confirmed through the XPS analysis. Figure 8.8 (a) reveals a survey spectrum of PPy. The survey spectrum for PPy has revealed the presence of O 1s, N 1s, and C 1s and a peak that corresponds to the elemental composi-tions of the ingredients constituting PPy. Figure 8.8 (b) depicts the core-level XPS spec-trum corresponding to O 1s, N 1s, and C 1s. The presence of large full-width at half maximum value in C-C bond has been revealed through the towering value of count

(s) in the core-level spectrum C 1s. The towering value of the count (s) corresponded to the binding energy of 284.60 eV. Deconvolution of the core-level spectrum of C 1s into two carbon neutral peaks corresponding to the binding energy of 286.68 eV has been revealed. This deconvolution can be attributed to –COH and –COOH, respectively. Figure 8.8 (c) reveals the spectrum corresponding to the N 1s in PPy. The presence of the neutral nitrogen (-NH-) in the pyrrole ring can be attributed to the N 1s core nitrogen peak. This has been revealed to be at the binding energy corresponding to 399.6 eV. Figure 8.8 (d) corresponds to the O 1s peak in the XPS spectrum of PPy. Deconvolution of the oxygen into three peaks has been observed and these peaks correspond to the binding energies at 530.44, 531.64, and 533.44 eV. The O = C bond can be attributed to the binding energy peak at 531.64 eV, while the C-O bonding corresponds to the energy peak at 533.44 eV. All the peaks corresponding to the binding energy indicate the PPy formation.

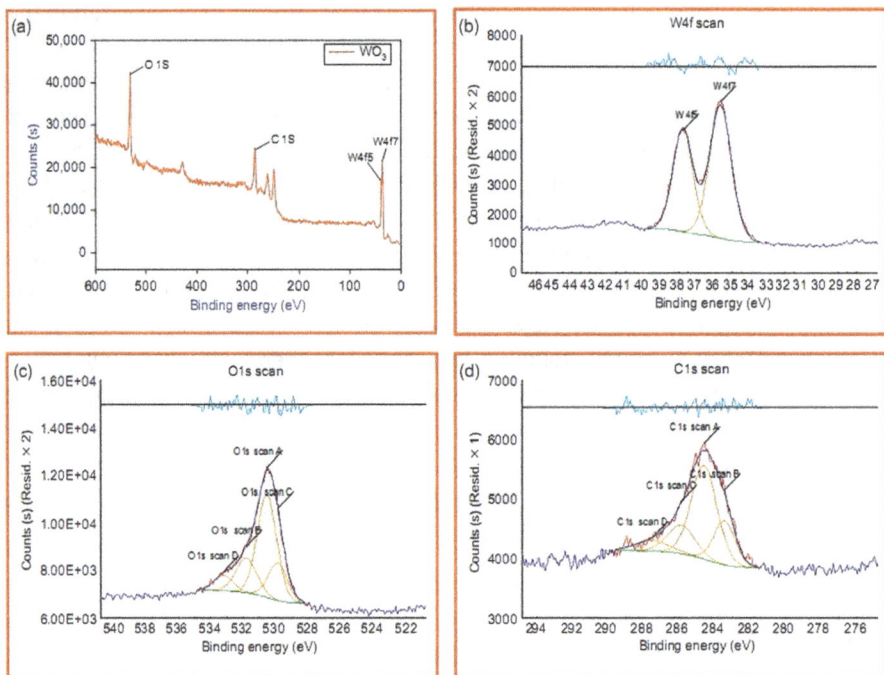

Fig. 8.8: Survey spectra of (a) WO_3, (b) C Is, (c) N Is, and (d) O Is (reproduced with permission from [56]).

8.5.2 XPS study of PPy-WO₃ nanocomposites

XPS has been employed to analyze the chemical states and composition of the surface for the PPy-WO₃ hybrid nanocomposites. Figure 8.9 (a) depicts the XPS spectrum ob-

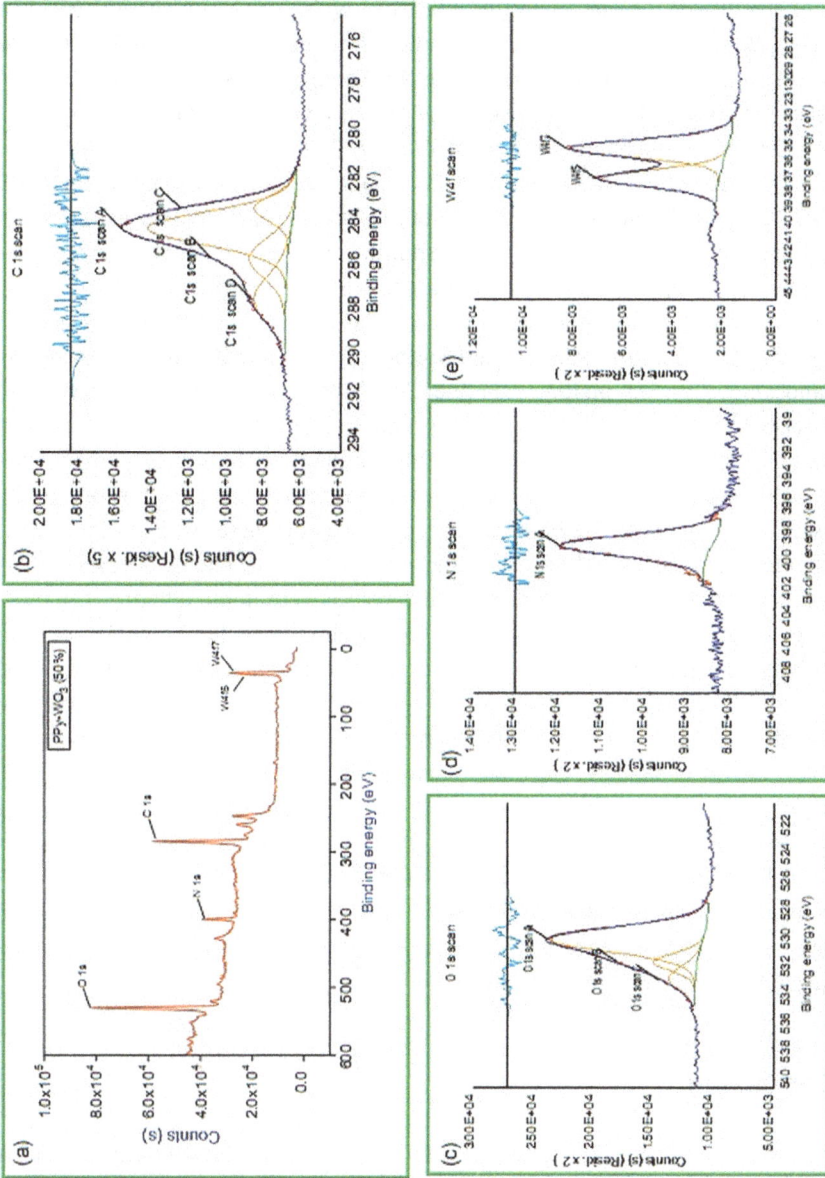

Fig. 8.9: Survey spectra of (a) PPy-WO₃ nanocomposite, (b) C 1s, (c) O 1s, (d) N 1s, and (e) W4f (reproduced with permission from [56]).

tained for the PPy-WO$_3$ nanocomposites fabricated with their equal percentages. The binding energy of 284.6 eV corresponds to the C-C bonding and the binding energy peak at 399.56 eV corresponds to that of the N-C bonding. These are the binding energy peaks for C 1s and N 1s, respectively. The presence of oxygen is reflected by the O 1s peak and this is at the binding energy of 530.7 eV. This also corresponds to the presence of O2_ in the lattice structure of PPy-WO$_3$ nanocomposites. The peaks for W4f5 and W4f7 correspond to the binding energy of 37.63 eV and 35.07 eV, respectively. The binding energy at 2.1 eV corresponds to the splitting of 4 f doublet of W. This observation reveals the energy position corresponding to W6+ oxidation state.

XPS analysis was also employed to confirm the core sheath structure associated with the PPy-WO$_3$ nanocomposites. Figure 8.9 (b–e) reveals the XPS spectra of the PPy-WO$_3$ nanocomposites. The binding energy of 284 eV corresponds to the characteristic peak associated with the carbon 1s peak. Asymmetric nature of the carbon peak has been revealed in the obtained XPS spectra. Using the shape analysis through the utilization of Gaussian-Lorenzian fitting functions, the asymmetric carbon peak can be decomposed into four lines. The binding energy of 283.6 eV corresponds to the lowest binding energy and the second peak corresponds to the binding energy at 284.3 eV. The aforementioned lines are recognized as *a* and *b*, respectively, in the PPy ring. The binding energy corresponding to 286 eV reflects on the third Gaussian-Lorenzian component. This may be attributed to the presence of doped contaminants and hydrocarbons. The fourth peak corresponds to the binding energy at 288 eV and is associated with the photon emission. Figure 8.9 (c) reveals the presence of oxygen and this corresponds to the binding energy of 529.8 eV. The three Gaussian functions can be fitted with the wide peak of O 1s spectrum and this corresponds to the binding energies of 530.1, 531.9, and 533.2 eV. These peaks reveal the presence of OW-O, OOH, and absorbed water, respectively. Figure 8.9 (d) reveals the deconvolution of the N 1s region and hence the presence of two nitrogen heteroatoms. The neutral N in the pyrrole ring is attributed to the binding energy at 399.7 eV. Figure 8.9 (e) reveals the peaks at 37.3 eV and 35.2 eV corresponding to W4f5 and W4f7, respectively.

8.6 Conclusion

XPS spectra for any material are obtained by irradiating its surface with the X-rays and measuring the kinetic energy (KE) as well the electrons escaping the material surface. The present chapter aims to provide an overview of the XPS analysis and hence aid the readers in understanding the atomic level behavior of the materials through XPS employability. XPS is employed to measure the elemental composition, and detect the contaminants, local bonding of atom, density associated with the electron states, binding energy of electron states, uniformity of elemental composition, etc. A few examples have also been considered for better understanding of the XPS methodology towards the end of the chapter.

Chapter 9
Ultraviolet photoelectron spectroscopy

9.1 Introduction

Ultraviolet photoelectron spectroscopy (UVPS) is an effective method in spectroscopy to identify the electronic states of atoms and molecules using He I and He II radiation sources. In UVPS, high-energy ultraviolet beam is incident on the sample atoms and molecules to ionize them, resulting in emission of low-energy photoelectrons. The ionization process can be given as:

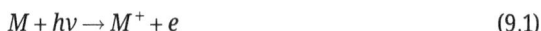

$$M + h\nu \longrightarrow M^+ + e \tag{9.1}$$

where M is the atom or molecule, $h\nu$ is the energy of the photon and e stands for the photoelectron. The kinetic energy of the photoelectrons (K.E.) emitted from the photoelectric effect under vacuum can be given in terms of the incoming photon energy and binding energy (B.E.) as:

$$K.E. = h\nu - B.E. \tag{9.2}$$

Here, the kinetic energy of the photoelectrons can be measured experimentally using a hemispherical energy analyzer while the photon energy from the monochromatic source is recorded. Thus, the B.E. can be found out from the above equation. UVPS system was first developed by Turner et al. [57] after which a number of atoms and molecules have been studied and experimented.

Monochromatic U.V. radiation and resolution are basically the major components, though the resolution required is less important due to the low-energy photoelectron emission in UVPS. Another advantage is that the intensity of the photoelectron released in an UVPS can be increased by the third power of the wavelength of the electromagnetic radiation, as a result of which greater photoelectron intensity can be obtained using UVPS. UVPS proves to be a better instrument as compared to the X-ray photoelectron spectroscopy in analyzing the atomic and molecular build of the gases, since the intensity can be elevated for better electron count rates of over 1000 counts per second.

One major setback of the UVPS [58, 59] instrument is its inability to study atoms and molecules in a bulk solid, and is only limited to studies of the sample surface. This is due to the smaller penetration of the UV photons and the smaller escape depth of the photoelectrons. In recent times, UVPS instrument has been widely used in studying gases in the form of adsorbent on the surface of the metal.

https://doi.org/10.1515/9783110997590-009

9.2 Instrumental details and working

As the ultraviolet radiation can be easily absorbed by the interacting species in the medium of its path, windows are generally not provided between the source and the target material. The components in the ultraviolet photoelectron spectroscopy can be mainly classified into

i.   Radiation source
ii. Vacuum system
iii. Monochromator
iv. Electron/ion analyzer and
v.   Particle detectors

9.2.1 Radiation source

Ultraviolet radiation source can be in the form of an atomic line radiation or high-flux photon radiation (using fast moving electrons circulating in a magnetic field). These are described below:

9.2.2 Atomic line radiation

Atomic line radiation is generally obtained by passing high voltage discharge through gases, mainly Helium, giving out radiation in the form of spectral lines. The discharge through the other gases produces a multilined spectral lines, which need to be filtered for a certain spectral line using a monochromator. This is tackled by using Helium that emits He I resonance radiation having a wavelength of 58.4 nm and energy of 21.22 eV. For the radiation to occur, a capillary discharge tube is employed where the gases at a certain pressure are excited by the application of a steady voltage producing a glow discharge.

A capillary discharge tube (Fig. 9.1) mainly consists of a quartz or boron nitride capillary, having a diameter of 0.1 cm and length of 3 cm, supported by the molybdenum electrodes through which the discharge is passed. The anode is basically provided with fins to avoid the complications in water cooling systems and can be conveniently cooled using air circulations over the fins. Desired gases are constantly pumped through the discharge to generate excitation in the atomic level, which gives out the line spectrum. High voltage of around 2 kV, with a combined resistor of 10 kohm, is applied under a gas pressure of 1 Torr. The current through the discharge can be adjusted by manipulating the voltage and the resistor. However, a higher voltage of around 3kV is required to initiate the discharge. The use of quartz capillary avoids the need for cooling, while the incoming gas needs to be purified of the impurity spectral lines by activated carbon, followed by nitrogen-cooled trap. Another radiation line at the wavelength of 30.4 nm and energy of 40.8 eV in pure helium excited

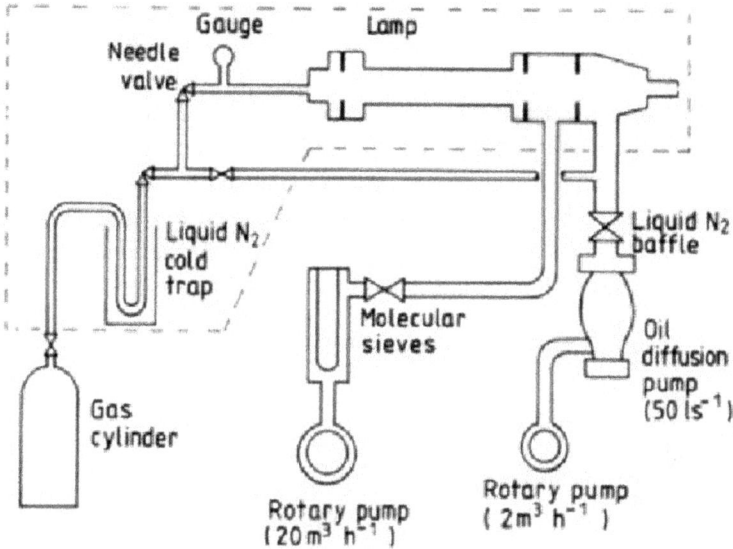

Fig. 9.1: Capillary discharge tube.
Source: A capillary discharge tube for the production of intense VUV resonance radiation (1983) (reproduced with permission from [60]).

capillary discharge tube produces He II resonance radiation. A sufficient intensity in this radiation type requires a low pressure of pure Helium and a high current density through the discharge, which is basically difficult to achieve due to the break in discharge at the low pressure of gases, and is tackled using a larger diameter cathode. The voltage supplied is generally 6kV, while a current of 100 mA is passed. The starting discharge is operated at a pressure of 1 Torr and is slowly reduced to 0.2 Torr, with increase in voltage through the ballast resistor. Cooling of the anode is necessary due to which fins are provided and a forced convection air system is installed.

Other spectral lines in the range of energy from 20 eV to 40 eV can also be produced using a capillary discharge tube as described above for a He II resonance radiation, but for more number of spectral lines of higher intensity, the duo plasmatron light source is the most appropriate one. Plasmatron is basically an ion source that is also an effective atomic line radiation source, as the generation of ions to produce higher intensity beam is very similar to the generation of the high intensity spectral lines involving lower energy state excitation of electrons from atoms/ions. The instrument consists of a heated filament involving electron excitations, which maintains the arc, and is confined by applying a magnetic field between the anode and the intermediate electrode. The operating gas pressure employed in Plasmatron lies in the range between 0.06 and 0.4 Torr and an arc current of 3A at a voltage of 150 V is applied.

9.2.3 Synchrotrons

Synchrotron (Fig. 9.2) was first used by Madden and Codling in 1963 to analyze the auto ionization states of He by photo absorption. These are basically fast moving charged particles, maintained in an orbit by strong magnetic field under a high voltage, which emits electromagnetic radiations. A high flux of photon radiation can be witnessed by electron synchrotrons due to the high current maintained and the high radial acceleration of the electron beam. The beam is of thickness less than 1 mm, which emits radiation in the wavelength range of 0.1 to 200 nm, and is highly polarized and pulsed.

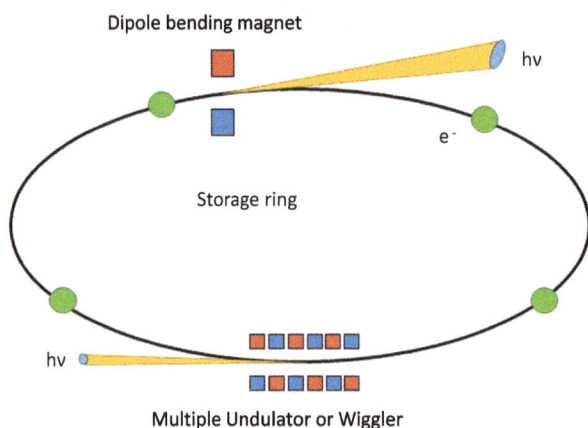

Fig. 9.2: Synchrotron.
Source: Schematic representation of the synchrotron radiation source (2016) (reproduced with permission from [61]).

9.2.4 Monochromators

Monochromators are generally used in cases where the light from the source needs to be dispersed for a better analysis of the spectrum characteristics, such as the small peaks and structures, intensities of the peaks, impurity and constituent detection, auto ionization phenomenon, etc. A number of disadvantages associated with the use of the monochromators to analyze samples using UV photoelectron spectroscopy must be consider by the spectroscopist to decide whether to use it or not in their experiments. These disadvantages can be summarized as the loss of intensity, capital and running costs, positioning and adjustments of the monochromator, deciding the small spatial extent of the beam that is to be observed, interference with the particle spectrometer, etc.

A variety of monochromators employed in UV photoelectron spectroscopy makes the selection of a monochromator for a specific problem and experiment difficult. Vacuum monochromators operate at a wavelength range of 10–100 nm, which consists of a diffraction grating that is coated with a transparent material with zero absorption. To avoid further absorptions, the source light must be passed through the least number of reflections or minimum number of mirrors. The angle of incidence must be maintained at a specific glancing angle due to an increase in the reflection coefficient, with increase in the angle of incident. The reflectors used in vacuum monochromators are made of gold and platinum that work well in the desired spectral region.

Fig. 9.3: Seya Namioka monochromators.
Source: Seya Namioka monochromator (retrieved on 16 February 2020), McPherson (reproduced with permission from [62]).

Monochromators with Seya Namioka mounting (Fig. 9.3) are the most simple and frequently used mountings. It is named after its inventor and is widely available commercially. The basic feature of this mounting involves a rotating diffraction grating onto which the light source coming in through the entry slit is incident at an angle and then collected using a fixed exit slit. A region of a particular wavelength of the spectrum is generally allowed to be collected through the exit slit and the whole setup is kept in a vacuum environment. The image obtained is elongated as compared to the line source and an optimum resolution of about 0.02 mm can be obtained. Astigmatism is a common problem that changes the shape of the light passing through the exit slit to a diffused focus beyond the target. This can be removed by employing a concave or cylindrical mirror placed between the entrance slit and the diffraction grating, which cancels out the astigmatism produced by the grating. Another solution for reducing the astigmatism is to use an ellipsoidal or toroidal shaped grating surface instead of a concave shape, since in a toroidal grating the radius of curvature is reduced in the vertical plane as compared to the radius of curvature in the horizontal plane where diffraction takes place. The incident light strikes the grating at an angle of 142°, and therefore shorter wavelength light performance is much improved.

Detection of the light from the monochromator is done using a coating of sodium salicylate over the glass disc, which produces a blue fluorescent light when irradiated with a weak beam of ultraviolet light. The output of the system is basically proportional to the intensity of the light and does not depend on the wavelength over a wide range.

9.2.5 Vacuum requirements

A vacuum environment is required for the successful operation of the electron spectroscopy, mass spectroscopy and electron multipliers, which varies with different circumstances. Electrons might suffer inelastic collisions by the background gas in an electron spectrometer on their path to the analyzer, resulting in intensity and energy loss at even a pressure of 10^{-5} Torr. Due to the involvement of a number of factors, a safe pressure range to conduct the experiment therefore cannot be overlooked. These factors include the length of the trajectory of the electrons, resolution requirement, proximity of adjacent peaks, scattering cross sections, etc.

For an ultraviolet photoelectron spectroscopy, generally a three-stage differential vacuum system is used, which consists of a 10 cm oil diffusion pump with trapped liquid nitrogen. The spectrometer chamber is generally kept under a vacuum pressure of 10^{-6} Torr and is made of stainless steel sheet with aluminum flanges of about 30 cm diameter. Helium discharge is generally maintained at a pressure of 1 Torr and a rotary pump with a capacity of 100 L/min is generally used to pump helium. The sample inlet-outlet system also employs an oil diffusion pump for vacuuming, having a diameter of 10 cm. The collision chamber is kept at a pressure of 1 Torr and based on all the above arrangement, the diameter of the He I photon source and the electron outlet orifices are adjusted. The commonly employed vacuum-measuring equipment include the thermocouple vacuum gauge and hot cathode ionization. For isolation from the earth's magnetic field, the spectrometer chamber is generally shielded with a thick micro metal sheet of 0.5 mm.

9.2.6 Energy analyzer

The photoelectrons emitted are analyzed at a high resolution using a 180° double hemispherical energy analyzer (Fig. 9.4), fabricated out of extruded aluminum rods. The accuracy of the surface of the hemispheres lies within ±0.025 and both the hemispheres are mounted on an aluminum plate, with slits being provided at 180° apart the entry and exit planes. The hemispheres and the aluminum plates are insulated using Teflon washers. A magnetic trim coil having a single turn is used around the outer hemisphere to fine-tune the signal.

Here, the electrons are passed through the entry slit of the analyzer at a zero potential and a certain kinetic energy, which are collimated by applying a voltage, V_p

across the hemispheres, until it reaches the exit at a zero potential again. The kinetic energy of the electron, K.E., can be expressed in relation to the voltage as given by:

$$K.E. = eV_p \left[\frac{1}{r_1} + \frac{1}{r_2} \right] \qquad (9.3)$$

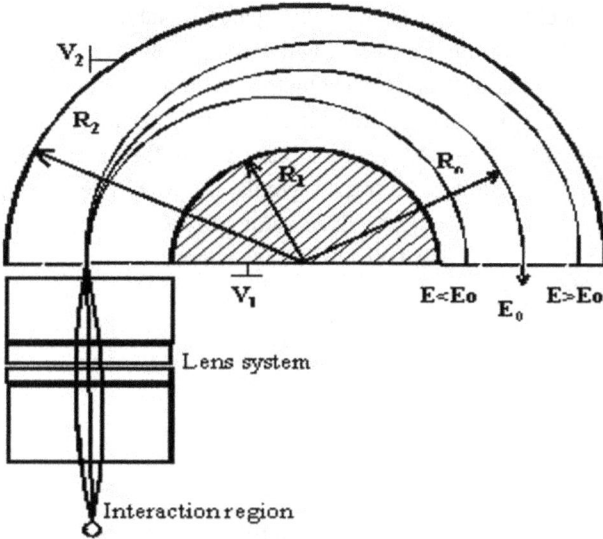

Fig. 9.4: Energy analyzer.
Source: HDA dispersion (retrieved on 16 February 2020), GCK (reproduced with permission from [63]).

where r_1 and r_2 are the outer radius of the inner hemisphere and the inner radius of the outer hemisphere, respectively. The resolution of the analyzer can be given as:

$$\frac{\Delta E}{K.E.} = \frac{S}{2r_0} \qquad (9.4)$$

where S is the slit width and $r_0 = (r_1 + r_2)/2$. With the probable range of binding energy of the electrons and hence kinetic energy ranging from 15.22 eV and 2.22 eV, and a suitable slit width of about 1 mm, r_0 can be evaluated, depending on the resolution required.

9.2.7 Electron detectors

Since the number of electrons received after the energy analyzer phase per second is very low, in the range of 10^3–10^5 electrons/sec, these electrons are multiplied in their number to about 10^8 electrons/n sec using channel electron multipliers as already been discussed in a different chapter. A voltage of about 2kV is applied across the channel electron multiplier, which accelerates the electrons to collide with the channel internal surface to produce numerous secondary electrons. The electrons are then detected to their final amplified signal, employing preamplifiers, rate meter, pulse shaping, counter, and recorder. The amplitude of the voltage obtained is dependent on the value of the resistance and the capacitance in the anode circuit of the amplifier. Commonly used amplifiers like the AMPTEK A101 PAD generally have a nominal threshold sensitivity of about 0.16 picocoulomb charge, which is equivalent to 10^6 electrons, and can be enhanced to 10 picocoulomb by varying the resistance. After amplification, the output is interfaced with a rate meter, which employs a decade counter to accurately count the charge pulses. The output is then passed on to the monostable where the pulses are fixed with a width of 30 microseconds and a voltage of 4 V, which is again changed by a voltage translator to a fixed value of 12 V. The pulses are transformed to an analogous signal by a frequency-to-voltage converter and then to the recorder via a buffer amplifier.

9.3 Sample analysis

9.3.1 Sample surface analysis

For a solid sample, the penetration of the ultraviolet source beam is influenced by the absorption of the photons by the atoms and the molecules below the sample surface. The photoelectrons produced on the surface of the solid are released into vacuum without any constraints, with a certain kinetic energy (e). While the photoelectrons released from below the surface collide on their way to the vacuum, with other atoms and molecules, knocking off other electrons, they exhibit a lower energy. The spectrum produced as a result of these electrons that suffered loss in energy while ejecting out into vacuum will contribute to the background of the spectrum. The depth in the solid sample up to which the electrons will be ejected in the vacuum without any collision or loss of energy is called the 'escape depth'. Any photoelectrons emitted from below the limit of the "escape depth" supposedly will knock off other electrons, thereby losing energy. It can be understood that with greater photoelectron energy, the possibility of the electrons losing energy is low and a greater "escape depth" can be achieved. This photoelectron energy is directly dependent on the excitation energy of the light source. Thus, it can be observed that in He I excited photoelectrons, while the "escape depth" extends to an order of a few 100 picometers, for an Mg $Mg\ K_{a_1 a_2}$ X-ray source, it extends to an order of two thousand to ten thousand picometers. For heavier atoms, the escape depth is higher due to higher collision and energy loss. While analyzing the bulk material, it must always be made sure that the compositions on the sample surface exactly match the composition within the sample.

Under high vacuum pressure, in the range of 10^{-6}–10^{-8} Torr, the water, hydrocarbon, and oxygen from the pump oil tend to condense on the surface of the sample of about 1 monolayer/ per second, which gives rise to additional features and peaks on the sample spectre. This condition can be put to advantage by employing it for calibration and it then also generates hurdles for surface studies. But, as the vacuum condition approaches to a minimum of 10^{-10} Torr, contamination is cleared on the surface of the sample and actual photoelectron spectrum is obtained.

Constant depletion of photoelectrons from the surface of the sample by energetic photons generally produces a positive charge on the surface. This buildup of positive charge on the surface also prevents the photoelectrons from escaping, thus reducing the kinetic energy of the particles. This condition can be tackled in the case of conducting samples by simply connecting the sample to the earth. The electrons will be replenished constantly, maintaining a neutral condition for more photoelectrons to be ejected without any constraints. In the case of insulators or nonconductors however, mere earthing the sample will not suffice and additional techniques must be employed. Secondary electrons are generally bombarded using an electron gun by collecting and diverting the electrons to the positive charge-induced surface, preventing loss of kinetic energy by the emitted photoelectrons. Due to the use of low-energy electrons for bombarding the surface, it helps in a self-replenishing mechanism without any control.

For calibration, a material is placed on the sample surface, whose fermi level is accurately known. It is electrically connected to the sample and does not react with the sample surface. Under such conditions, the calibrant material and the sample material will share an equilibrium fermi level, with both having the same surface charge. Features that are commonly employed for calibration include C1's peak of the hydrocarbons being adsorbed on the surface of the sample from the pump oil, the Au $4f_{7/2}$ peaks of the metallic gold, which can be sprinkled in the form of powders on the sample surface, and various other peaks of the calibrants internally mixed with the sample. Though a number of problems arise during the calibration for a specific experiment, the equilibrium assumption in the fermi levels of both the sample and the calibrant must be the best approximate. For gaseous samples, the calibration is generally a simple process due to the reduction in the effects of the contact potential and the space charges between the spectrometer and the sample.

9.3.2 Photoelectron intensities

The intensities of photoelectrons can be basically represented by the area under the peak on the spectrum produced, which is proportional to the number of photoelectrons released from the atoms (or the number of atoms in the molecule). Assuming a homogeneous concentration of atoms and molecules (C) in the sample system, with no collisions suffered by the photoelectrons from the sample atom to the spectrometer, the intensity (I) can be given as:

$$I = I_0 aCkx \qquad (9.5)$$

where I_0 is the intensity of the incident photon beam, k is an instrument factor for a specified instrumentation and energy of the incident photon, x is the length of the sample studied, and a is the crosssection for photo-ionization.

The instrument factor (k) basically depends on the design of the instrument used, which influences the intensity of the spectrum observed. A design involving high-intensity sources, and a high sensitivity to different kinetic energies and sensitive detection systems will yield a higher intensity and accuracy in quantifying the interpretation.

The photoelectron crosssection (a) can be related to the absorption characteristics in a photoelectron excitation phenomenon. When a beam of high-energy photons such as the far u.v. and X-ray beam is irradiated on a sample, the radiation that is transmitted (I') has an intensity lower than the intensity of the initial photon beam (I_0), where both can be related as given by Beer's law:

$$I' = I_0 e^{-\mu \rho x} \qquad (9.6)$$

where μ is the mass absorption coefficient of the radiation and ρ is the density of the sample material. The mass absorption coefficient is related to the absorption of the electromagnetic radiation by the sample, which basically can be separated into two types: the mass photoelectric absorption coefficient and the apparent mass scattering absorption coefficient. The mass photoelectric absorption coefficient (a) represents the absorption associated with photoelectron emission when a high-energy photon beam loses its energy completely. The apparent mass scattering absorption coefficient (b) is due to the scattering effect of the high-energy light source, which is dependent on the wavelength of the light source, the atomic number, and the mass of the sample material. As the wavelength of the light source decreases from the far ultraviolet to the X-rays, scattered photons deflect out from the path of the detector, thereby losing intensity. Therefore, mass absorption coefficient can be given as:

$$\mu = a + b \qquad (9.7)$$

the mass absorption coefficients (a and b) generally are dependent on the atomic number (Z) of the atoms in the sample and the wavelength of the incident light (λ), which are given by:

$$a = KZ^4 \lambda^3 \qquad (9.8)$$

$$b = K'Z^p \lambda^q \qquad (9.9)$$

where K and K' are constants, "p" takes the values from 1 to 2, and q is any number less than 1. The contribution of "b" is generally neglected except in the case of small wavelength values and small atomic numbers. In such conditions, the mass absorption coefficient will be equivalent to only the mass photoelectric absorption coefficient, which in turn is proportional to the photoelectron cross section (a) as given below:

$$\mu \approx a = a(L/A) \tag{9.10}$$

where L is the Avogadro's constant and A is the atomic weight of the sample atoms. Thus, it can be seen that with a higher atomic number, higher weight of the atoms, and longer wavelength of the incident radiation, a higher value of the photoelectric crosssection can be obtained. Hence, ultraviolet radiation can produce a higher intensity on its spectrum as compared to the X-ray excitation.

From the above eq. (9.5), it is clear that with increase in the concentration of the sample atoms and molecules, the intensity of the photoelectron emission increases, though this theory only applies appropriately to the gaseous sample. For a solid sample, when the concentrations of atoms and molecules are increased, it is very unlikely that the photoelectrons emitted from the surface atom will collide with another atom or molecule. Also, as seen in the solid sample surface analysis section, the photoelectrons emitted from below the surface might collide with the surrounding atoms and molecules, resulting in a loss of energy and intensity. Therefore, the above eq. (9.5) has to be modified with the assumption of a homogeneous solid sample, to examine the appreciable number of collision between the photons and the atoms, which is given as:

$$I = FaDkd\left(1 - e^{\frac{-x}{d}}\right) \tag{9.11}$$

where D is the density of the atoms, x is the effective thickness of the solid sample, depending on the angle of release of the photoelectrons from the surface and d is the escape depth.

9.3.3 UV spectra and their interpretation

UV photoelectron spectra of the valence orbital can be interpreted in comparison to the molecular orbital energies using approximation, where a number of cases have been identified with the Koopman's theorem-based calculations not fitting the observed spectrum. There always remain some proportions of organic molecules with various complexities, which can be considered for interpreting the valence orbital spectra.

During the analysis of the output spectrum from a UV photoelectron spectroscopy, a number of features such as the vibrational fine structure, instantly give information regarding the types of molecular orbitals from which the photoelectrons are ejected. Considering a number of related compounds, their UV photoelectron spectrum can be empirically co-related for characterizing the chemical changes and changes in the binding energies of the obtained bands.

For gases, the U.V. photoelectron spectroscopy provides a high intensity spectre, which enables the detection of the transient species. Free radicals such as CS, SO_3F, CH_3, NF_2, and N_2H_2 are prepared in a flow system, where these gases are pyrolyzed or photolyzed before entry into the sample chamber under vacuum, to maintain a steady state concentration at the photoionization point. An in situ mass analysis is carried out of the ions produced by the photoionization process to monitor the undesired phenomenon of the decomposition of the gases and other impurity complications during the study of the photoelectron spectrum. A pulsed ion accelerating field is generated within the ionization region to perform mass analysis using a reversal electrostatic field and measuring the time of flight of the ions ejected out from the point of photoionization. Few other structures in the ultraviolet photoelectron spectra are described below:

Auto-ionization process can be a setback in the analysis of the uv photoelectron spectrum as it enhances the electron counts in the detector, leading to higher intensities. This basically depends on the velocity or the energy of the photon particles bombarded onto the sample surface. The higher the photon energy, the probability of auto-ionization decreases, since the interaction time between the photons and the electrons in the energy levels of an atom, reduces.

Auto-ionization is basically a process where neutral atoms and molecules are excited by the bombardment of photons of the correct energy by knocking out electrons from the lower energy state and transferring electrons from the higher energy state to the vacancy created in the lower energy state. Appreciable interactions between the ground state molecules and the excited state can be witnessed during the auto-ionization process as the probability of auto-ionization increases as compared to direct ionization. This affects the vibrational structure in the photoelectron spectrum and also enhances the cross-section of the emitted photoelectron, which results in an increase in intensity.

Predissociation of molecular ions during the photoelectron process is a common phenomenon in u.v. photoelectron spectroscopy. Predissociation eliminates the vibration of the fine structures of the photoelectron spectrum. Another phenomenon, named the "Jahn-Teller effect", is one where irregularly spaced vibrational structures are generated in the photoelectron spectrum of nonlinear molecules.

9.4 Examples on UV-Vis spectrum

9.4.1 Metallic nanoparticles as nanofillers

Free electrons within metals are referred to as plasmons, when they are quantized. These free electron lead to the generation of surface plasmons resonance (SPR) when they interact with light. The associated oscillation frequency for noble metals such as copper, silver, gold, and other prominent alloys is basically in the visible region. Therefore, the typical color of each of the metallic clusters is revealed at the nanoscale. The shape and size of the nanoparticles are the determinant factors for the oscillation modes of the surface plasmons. The interaction of the spherical nanoparticles with light reflects the isotropic behavior, and this is observed as a single SPR peak within the obtained UV-Vis spectrum. However, two major peaks are observed in the case of anisotropic particles. The first peak is the transverse peak, which is independent of the aspect ratio and the second is the longitudinal peak, which is dependent on the aspect ratio. These are reflected in Figs. 9.5–9.7. Valuable information with regard to these can be sourced from the absence /presence of the SPR peaks, the intensity of the peaks, and the widths of the SPR peak. UV-Vis spectroscopy becomes quite useful to understand the size and dispersity of the nanoparticles within polymeric nanocomposites.

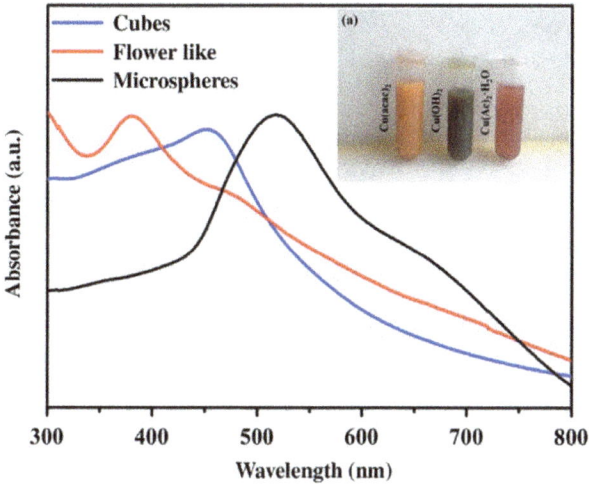

Fig. 9.5: SPR absorption peaks for copper nanoparticles (reproduced with permission from [65]).

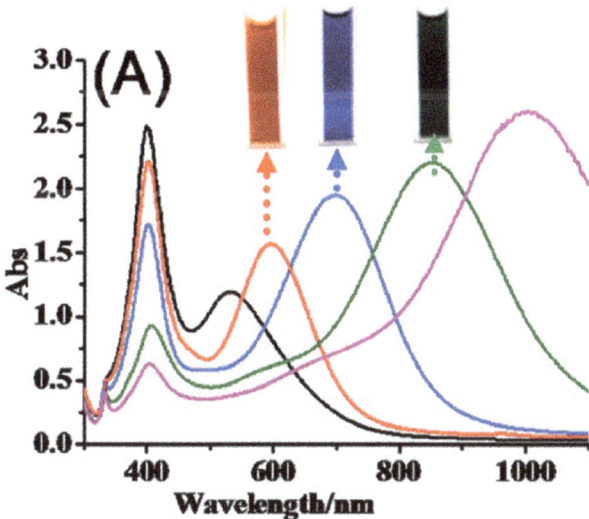

Fig. 9.6: SPR absorption peaks for silver nanoparticles (reproduced with permission from [66]).

Polyacrylamide-based nanocomposites containing homogeneously distributed metallic nanoparticles were fabricated by Zhu and Zhu [64].

This was accomplished through the use of microwave heating. UV-Vis spectroscopy was adopted to investigate the stability of the prepared nanocomposites' colloidal solution.

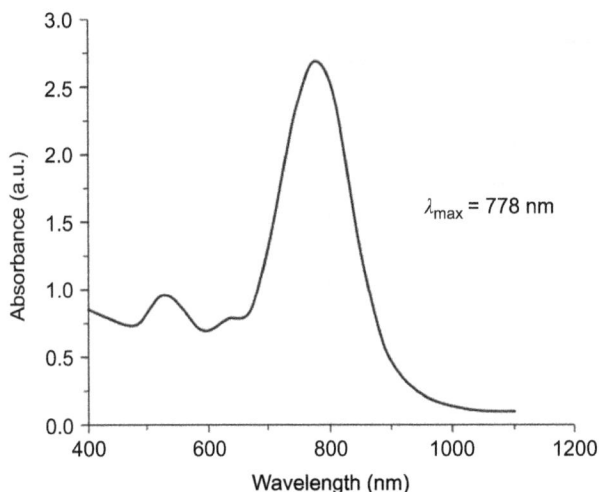

Fig. 9.7: SPR absorption peaks for gold nanoparticles (reproduced with permission from [67]).

The maximum SPR absorption is characteristically revealed by the silver nanoparticles at around 410–420 nm. The peak, in the case of the fabricated nanocomposites, has been revealed to be at 423 nm, indicating the presence of silver nanoparticles within the polymeric matrix. UV-Vis spectra was also investigated for the fabricated sample after storing for around eight months. No shift in the absorption band of around 423 nm was revealed in both the samples. This indicated the higher stability of the fabricated nanocomposites. The UV-Vis spectra has been depicted in Fig. 9.8.

In another study by Khanna et al. [12], the UV-Vis spectra was employed to investigate the silver nanoparticles nanocomposites with poly(vinyl alcohol) matrix. The quality of the embedded nanoparticles was investigated through the UV-Vis spectroscopy, the nanoparticles being formed by the aid of two reducing agents. The SPR peak has been revealed to be present at 418 nm. This indicated the formation of silver nanoparticles within the polymeric matrix. The absorption peak has been revealed to be sharp for the silver nanoparticles formed using SFS (Fig. 9.9 (a)). On the other hand, the peak is broader in case the HF solution is used for the formation of silver nanoparticles (Fig. 9.9 (b)). This indicates an uneven size distribution of the silver nanoparticles.

The peak revealed at one-half the maxima is considered to be a very crucial parameter. In the study conducted, this was employed to comprehend the size of the particles and hence their distribution within the matrix and the colloidal suspension. This peak has been revealed to be at 110 nm in the case of the silver nanoparticles prepared using

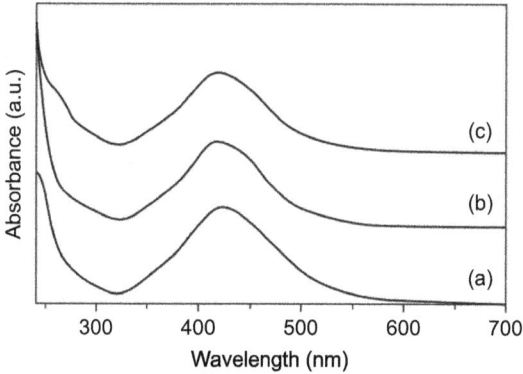

Fig. 9.8: UV-Vis absorption spectra of fabricated silver nanoparticles polymeric nanocomposites (reproduced with permission from [64]).

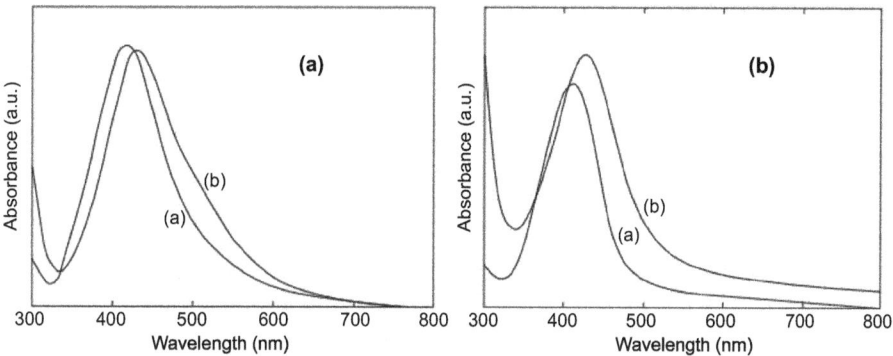

Fig. 9.9: UV-Vis absorption spectra of Ag/PVA nanocomposite formed using (a) SFS; (b) HF (reproduced with permission from [68]).

HH, whereas it was at 85 nm in case the nanocomposites were fabricated using the SFS solution.

UV-Vis spectroscopy was employed to understand the interaction between silver nanoparticles preformed in polypyrrole, carboxymethyl cellulose, and PVA [13]. The interactions were studied by following the SPR absorption peak. The SPR maximum for the silver nanoparticles in the aforementioned solution has been revealed to be at 417 nm (Fig. 9.10 (a). However, there is no peak in the case of the pyrrole solution, as can be observed from Fig. 9.10 (b).

The SPR absorption patterns for the PVA/Ag nanocomposite films have also been characterized for various percentages of Ag and some interesting results have been obtained. A broad surface plasmon band for absorption can be observed in Fig. 9.10 (c) and the peak corresponds to 420 nm. The agglomeration of the Ag nanoparticles can be one of the reasons for the broadening of the peak and its shifting to a longer wavelength.

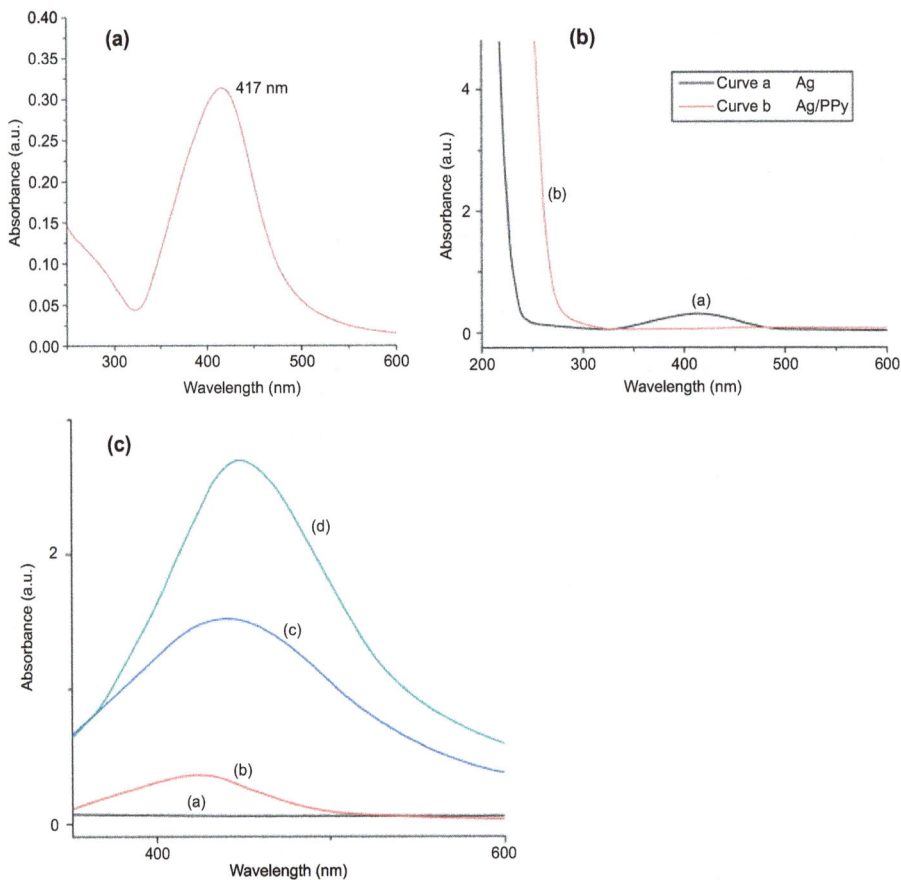

Fig. 9.10: UV-Vis absorption spectra of the colloidal solution of Ag nanoparticles in (a) polypyrrole; (b) carboxymethyl cellulose; (c) PVA (reproduced with permission from [69]).

A novel method for the co-deposition of Teflon with different nanoparticles such as Au, Ag, and Cu was proposed by Biswas et al. [70]. The developed nanocomposites have been characterized using UV-Vis spectroscopy. Tunable multiple SPR responses have been observed from such systems at different UV-Vis wavelengths. A double wavelength plasmon response has been revealed in the spectrum of Teflon Ag/Au nanocomposites (Fig. 9.11). The lower wavelength corresponds to the Ag nanoparticles while on the other hand, the Au particles are reflected at the higher wavelength.

9.5 Conclusion

The present chapter has provided an overview of the Ultraviolet photoelectron spectroscopy (UVPS), which is an effective method in spectroscopy to identify the elec-

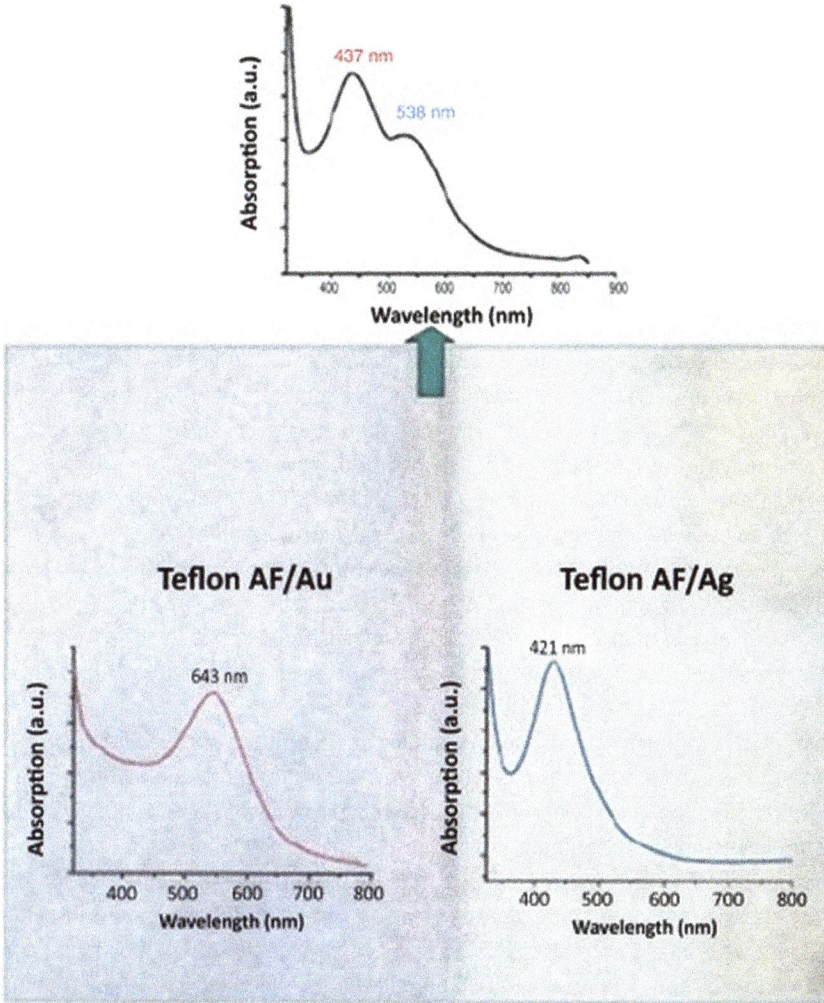

Fig. 9.11: Particle plasmon resonances of Teflon AF/Ag and Teflon AF/Au nanocomposites (reproduced with permission from [70]).

tronic states of atoms and molecules. The details with regard to the instruments employed have been elaborated that can aid early beginners in easily comprehending the working of the instrument and its associated components. The principles associated with the sample analysis have also been depicted so that the basics associated with sample handling and data from the analysis can be comprehended. In the end, few examples on the analysis of polymeric composites have been discussed.

Chapter 10
Fluorescence spectroscopy

10.1 Introduction

The phenomenon of fluorescence and phosphorescence due to various transitions in the electronic and vibrational states can be suitably presented in the form of a Jablonski diagram. The electrons occupying the molecular orbitals around the nucleus forms the skeleton of any study on polyatomic molecules. The electrons posses few properties in their orbitals; they are its angular momentum and energy arising due to their motion around the nuclei, in addition to its spin about its own axis. Based on the Pauli's exclusion principle, it is known that only two electrons can be accommodated in any non-degenerate molecular orbitals, with both possessing equal and opposite spin angular momentum about their axis (–1/2 and ½ spins), cancelling out to give a total of zero angular momentum. The electronic configuration obtained in such conditions (ground-state conditions) can be represented in terms of multiplicity, which is basically the total angular momentum of the electron spin in the orbitals, including both the lower and higher electron-filled molecular orbitals, given by $2S+1$, where S is the total spin angular momentum, thus yielding a state multiplicity of unity, and are referred to as singlet states.

There exist a number of unoccupied orbitals along with the occupied orbital ones in the ground electronic state, and electromagnetic radiation (visible or ultraviolet) can be utilized for the absorption of energy by the electrons in the occupied orbitals and getting transferred to the unoccupied ones, as the first excited singlet state. This energy absorption by the ground-state electrons is basically due to the interaction between the electric field components of the electromagnetic radiation and the oscillating electrons within the orbitals. This process is also referred to as electric dipole transition and no change in spin angular momentum of the electrons is seen during this transition.

As the Pauli's exclusion principle does not apply to this state of molecular orbitals where the electrons are excited, they are free to alter their orientations of their spin, resulting in electronically excited states, possessing a multiplicity of 3 (total spin angular momentum of 1). This state can be referred to as the triplet state, which is lower in energy compared to the first excited singlet state. This is due to the repulsive energies of the electrons in the singlet and triplet states.

Fluorescence is a phenomenon generally arising when electrons in their excited states do not alter their spin states during emission transition. While, when the spin inversion is necessary for transition from the triplet excited state to the singlet ground state, the phenomenon of phosphorescence is generally exhibited. In this chapter, a general description of the instrument for fluorescence spectroscopy is pro-

https://doi.org/10.1515/9783110997590-010

vided. Various sample handling techniques, sample holders, and spectral analysis in the fluorescence spectroscopy have also been discussed in detail.

10.2 Instrumental details

Spectrofluorimeter (Fig. 10.1) is commonly used for distinguishing long-lived (greater than 1 ms) and short-lived emissions of a fluorescence and phosphorescence spectra. The basic components of a spectrofluorimeter can be described as below.

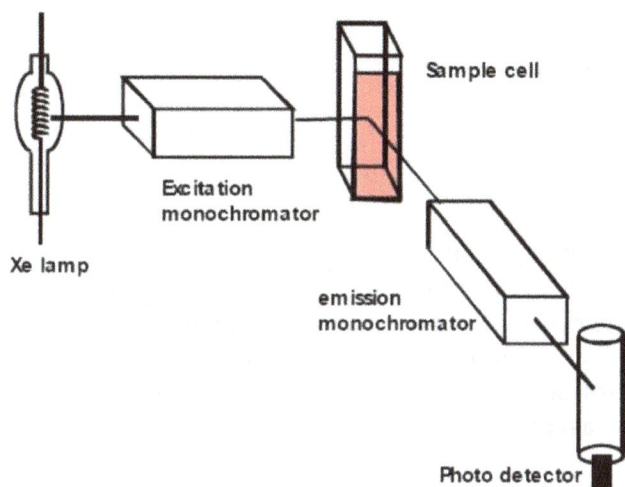

Fig. 10.1: Spectrofluorometer.
Source: Schematic illustration of the spectrofluorometer (2015), Researchgate [71].

10.2.1 Light source

Xenon arc lamp (Fig. 10.2) of power, ranging from 150 to 500 W, and run by a DC stabilized power supply is commonly used as a light source in a spectrofluorimeter. Radiation with wavelength in the range between 200 and 800 nm can be obtained from such a light source.

Xenon arc lamps contain a highly pressurized quartz-enveloped chamber, inside of which an arc is generated between two tungsten electrode poles using DC supply voltage. The xenon gas pressure maintained in the chamber is about 20 atm, with two tungsten poles separated by a distance of 1 mm to 1 cm, achieving a radiation wavelength of about 200 nm – these are few characteristics of the xenon arc lamp.

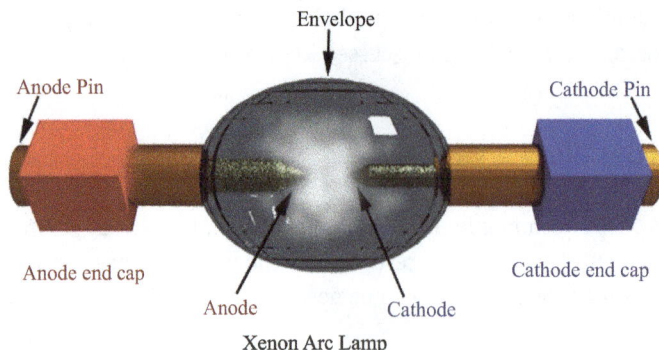

Fig. 10.2: Xenon arc lamp.
Source: Xenon arc lamp (2018), Electrical 4 U [72].

10.2.2 Monochromator

Dual monochromators are used in a spectrofluorimeter to select the radiation of a particular range of wavelengths, one as excitation and another as emission radiation selector. Generally, gratings of high resolution as monochromators, driven by a motor system, and an emission monochromator with better spectral analysis characteristics of luminescence as compared to the excitation monochromator are used. The fundamental workings of monochromators have already been discussed in other chapters.

10.2.3 Flexible cell housing

Flexible cell housing is used to mount a variety of fused silica cells with accommodation for both low and high temperature control systems.

10.2.3.1 Detection system
For detection of the output signal, a sensitive photomultiplier (Fig. 10.3) that multiplies the number of detected electrons to amplify the signal is used, which is then displayed on a strip chart recorder. To enhance the sensitivity, in some cases, single photon counting detection may also be employed.

10.2.3.2 Correction system
The spectral diagram obtained from the detection system is usually distorted in their true shape due to the dependencies of the sensitivity of the spectrofluorimeter and transmission deviations (through the optical devices) on the nonlinear behavior of the wavelength of the photon emission. The correction is done by conducting the experi-

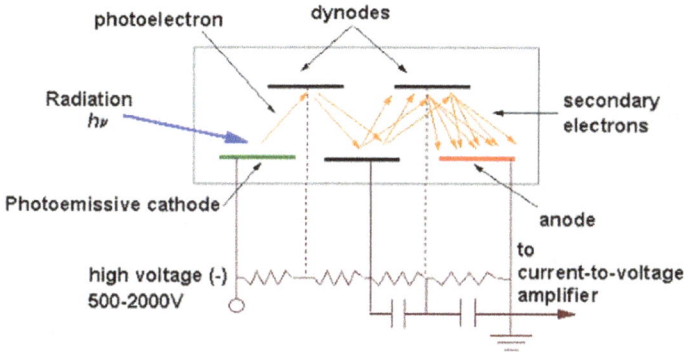

Fig. 10.3: Photomultiplier.
Source: Photomultiplier Tube PMT (1996), Science Hypermedia [73].

ment for a standard sample, of which the characteristics are known, and comparing the instrumental response error. This is then followed by experiments on the sample that is to be characterized.

10.2.3.3 Polarizer

For experiments that require polarized radiation transmission, a prism polarizer (Fig. 10.4) is generally inserted into the exit slit of the excitation monochromator. This polarizer, along with the analyzing polarizer, which is basically the same polarizer but inserted into the emission side slit of the emission monochromator, can measure the electric field vector transmitted parallel and perpendicular to that of the radiation source.

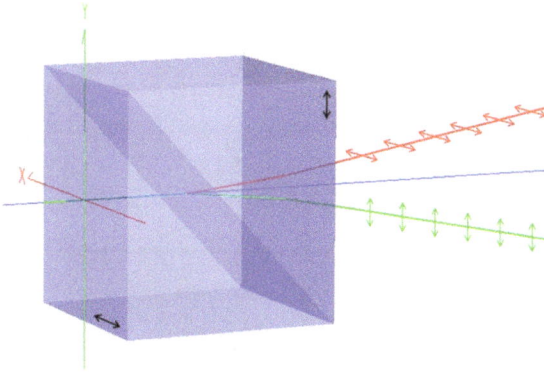

Fig. 10.4: Prism polarizer.
Source: Wollaston prism polarizer (2011), Photon engineering [74].

10.2.4 Phosphorimeter attachment

Phosphorimeter attachments are generally used for measuring the delayed emission time by interrupting the excitation beam periodically using mechanical and electronic devices. As the excited radiation is interrupted and allowed to be viewed by the detector only after a specific interval of time, the short-lived emission gets decayed to zero intensity and only the long-lived emission persists. Thus, it can also be employed to measure the decay time of luminescence. For mechanical systems such as rotating toothed chopper, the minimum decay time that can be measured (or measuring the lifetime of the short-lived emission) in the order of 1 ms. However, by using electronic devices, decay times of much shorter order of time can be measured.

10.3 Experimental sample handling and holders

Samples that are mostly used in fluorescence spectroscopy are in the form of solutions, with concentrated ones giving more precise and stable output than the diluted ones. Samples generally are contained in a cuvette (Fig. 10.5) or flowcell. The material for the cuvette is generally transparent to light wavelength; it is of major importance in a spectrofluorimeter and is available in shapes of square, round, and rectangular. Square cuvettes are the most widely used due to ease of determination of parameters such as path lengths and parallelism, and to maintain these during manufacturing. Rectangular cuvettes are rarely used while circular cuvettes are less expensive and are suitable to be used for routine applications.

Fig. 10.5: Cuvettes.
Source: 1 ml and 3 ml cuvettes (2016), Wikipedia [75].

Since fluorescence is radiated in all directions with equal intensity, the collection of the light can be made from either the front of the cuvette square, holding the sample, from the sample, at right angle to the incident beam, or in line with the incident beam. The collection system or the direction from which the fluorescence is to be collected depends on the characteristics of the sample. For a dilute solution sample, fluorescence is emitted in all directions equally, and therefore 90° collection arrangements are common to track fluorescence that occurs along the path of the incident beam through the sample. A 90° collection arrangements are employed as they minimize the effect of scattering of the fluorescent light by the solution and the cell. It can be observed that only a small fraction of the fluorescent light is actually collected (equal intensity from all directions) by the detector from a smaller volume of the sample, with no contribution from a large volume of the remaining solution. To minimize this inefficiency, microcells are used, having a dimension that takes into account the optical considerations of the instrument.

The absorbance of the sample solution plays a vital role in deciding the collection direction of the fluorescent light. As the fluorescence progressively distorts during the penetration through the bulk solution, the light collecting directions such as the 90° light collection and in-line with the incident radiation becomes fruitless, and only the front view of the cuvette collecting method is utilized though scattering from the cuvette wall, which will be large in such a case.

The concentration of the sample solution tends to have a direct effect on the absorbance of the different samples and there is a high probability that the sample solutions with high concentration (or absorbance) will be discarded for displaying a lower fluorescence effect. The absorbance of an unknown solution therefore should always be determined before any experiment and the concentration of the sample adjusted so that the absorbance is no more than 0.1 A.

The overall dimensions of the cuvette do not matter much in a spectrofluorimeter experiment due to the role of only a small central region on the cuvette in transmitting the actual source radiation. Also, precision remains intact, irrespective of the use of inexpensive test tubes or precision cuvette for sample loading. But variations in the width of the cell wall with respect to the subsequent cells walls might distort the final spectrum obtained if laboratory grade glass tubes are used with other light sources, in the case of repeated transmission. The native fluorescence of the glass materials will diminish the accuracy of the spectra and produce large blank values.

The glassware and cells used in spectrofluorimeter experiments must be thoroughly cleaned before use, by boiling in a container of 50% nitric acid and then rinsing with distilled water. Glassware can be cleaned in synthetic detergent solutions, and is generally preferred in laboratories. Though for doing so, first the fluorescence characteristics of the dilute solution of the detergent have to be measured at the analytical wavelength, and minimized.

Solvents and reagents must be checked periodically for any contaminants that can raise the blank values, and the common way to reduce the blank values to reasonable limits is by distillation and washing with acid, water, and bases. Water, in its deionized and distilled condition, is generally perfect to be employed as solvents. Few other precautionary measures while considering blank values of solvents and reagents can be given as follows:

1. Solvents must never be stored in glass containers as plasticizers and organic additives leaching can produce high blank values.
2. Reagents of high-quality grade must be used and must be controlled while using it on assay.
3. High blank values should create suspicion over the type of reagents and solvents used.
4. For calibration, the blank values of the reagents must be predetermined through analytical procedures before it is set to zero.
5. Fluorescence spectra obtained from contaminants in the solution should not be confused with the Raman spectra of the solution.

10.3.1 Working with dilute solutions

As dilute solutions are not so stable, it is a common practice to store the sample in the form of concentrated stock solutions and then dilute it to working standards. It is worth noting that the final diluted solution must not be stored for a longer period before employing it for florescence study.

Adsorption is a problem in spectrofluorimeters as organic substance tends to get adsorbed on the walls of the container. To avoid this, small amount of polar solvents are generally added to the nonpolar solution while measuring aromatic substances. Also, glassware must be thoroughly cleaned in acids.

Photodecomposition is a common phenomenon while irradiating the sample with intense light source in fluorimeters. Usually, a shutter system is employed to allow the intense light to penetrate the sample only for a brief period of time so that, to a high degree, photofluorescence can be prevented. After the shutter shuts down, the unaffected fresh molecules in the path of the previously illuminated light path absorb the energy to fluoresce. Photofluorescence can be completely avoided by altering the choice of the excitation wavelength and by diminishing the intensity using a narrow width slip or a neutral density filter.

Fluorescence intensity can also be affected (reduced) by the presence of oxidizing agents in solution such as dissolved oxygen and peroxides.

10.4 Spectral analysis and interpretation of fluorescence

Before going for the spectral analysis of the fluorescence phenomenon, certain characteristic properties and observables have to be thoroughly studied and kept in mind during the characterization of the different atoms and molecules. Observables include the intensity, position, and lifetime measurements that are actually observed, and are interpreted to give sensible conclusions. Fluorescence or emission of photons is triggered when a high-energy photon beam transfers its energy to the ground-state electrons, which absorb to reach an excited state and transition back to their minimum energy ground state. The basic properties of fluorescence can be described as below:

1. Emission energy (or fluorescence) is always lower than the absorption energy. This is due to the fact that the absorption energy is consumed in a number of ways (phenomena) other than by emitting fixed photon energy (de-excitation) of specific wavelength. These other phenomena that contribute to the fluorophore de-excitation include phosphorescence, photo-oxidation, radiationless losses, and energy transfer. The weaker these phenomena, the higher is the fluorescence contribution to flurophore de-excitation.
2. Phosphorescence energy is always lower than the energy of fluorescence.
3. Emission from an excited state to the ground state is totally independent of the excitation energy and wavelength.
4. There is no for of spin re-orientation in the absorption and fluorescence phenomena, though phosphorescence is affected by it. This also makes fluorescence a faster phenomenon than phosphorescence. Absorption occurs in a time order of 10^{-15} s while fluorescence lifetime lasts for a time order of 10^{-9} to 10^{-12} s. Phosphorescence takes a longer lasting time, ranging from milliseconds to hours.
5. The transition energy (E) of electrons between two energy states is the energy difference between the states and can be given in the form of energy of the photon absorbed as:

$$E = h\nu$$

   where h is the Planck's constant and ν is the frequency of light. Thus, it can be implied that the transition will only take place at a specific wavelength of light and an emission line should be observed in the spectrum. But a spectrum is usually seen instead of line due to the contributions of the vibrational and rotational states to the absorption and de-excitation energy.
6. Absorption is mainly concerned with the ground-state electronic distribution and characterizes it accordingly, whereas in fluorescence and phosphorescence, the electronic distribution of the excited state is what matters and is termed as "mirror image" of the excited state. Thus, a change in the fluorescence spectrum is observed when charged transfer between electrons occurs.
7. Temperature variations are important factors to be considered during a study on the fluorescent spectrum, which induces local and global changes in the motion of the fluorescent environment and the fluorescent.
8. The fluorescence spectrum is basically a plot between the intensity of fluorescence and the wavelength.

9. Fluorescence emission spectrum can be recorded by fixing the wavelength of the excitation source and varying the emission monochromator range. Similarly, the excitation spectrum can be recorded by fixing the emission wavelength and varying the excitation monochromator output.

10. As the absorption energy of the ground-state electrons is always greater than the emission energy of electrons in the excited state, the wavelength of the emitted photon (fluorescence) is always larger than the incident photon. This phenomenon is also known as the Stoke's shift. The spectrum obtained from fluorescence, if it contains two or more bands then the difference between the two most intense bands, is called the Stoke's shift.

10.4.1 Inner filter effect

The intensity of fluorescence spectra is directly proportional to the number of excited electrons taking part in the emission of photons and the optical density. However, in normal cases, the light entering the cuvette travels one-half of its path length before the emission of photons from the sample solution placed at the central region and another one-half before reaching the detectors. Due to such conditions, the intensity of the detected emission spectrum is generally taken to be the result of the optical densities at both the excitation and emission wavelengths. This auto-absorption by the solution decreases the intensity of the observed fluorescence spectrum and shifts the maximum emission band, which is generally referred to as the inner filter effect.

10.4.2 Fluorescence excitation spectrum

Fluorescence excitation spectrum is basically obtained by fixing the emission wavelength and varying the wavelength of the excitation light by the emission monochromators, which characterizes the electronic state of the ground-state electrons. As the excitation spectrum is generally disturbed by the variations in the light intensity of the excitation source and the intensity of the emission, both of which are dependent on the wavelengths, a reference substance for correction is generally adopted, such as the use of Rhodamine B, dissolved in glycerol.

Excitation of the Rhodamine B generally yields fluorescence excitation spectrum, which is linearly proportional to the intensity of the excitation radiation, with no dependency on the wavelength of the radiation, and thus characterizes the excitation lamp spectrum. An automated process is carried out inside the spectrofluorimeters where the actual fluorescence excitation spectrum is obtained, by dividing the recorded excitation spectrum with the excitation spectrum from Rhodamine B.

10.4.3 Mirror image rule

The emission spectrum obtained is basically the mirror image of its absorption spectrum only when the probability of the transition from S_1 (excited state) to the S_0 (ground state) is identical with the probability of the transitions from the S_0 to the S_1 states. This phenomenon is observed due to the weak interaction of the excited states of fluorophore with the solvent.

But in many other cases, the excitation of fluorophore triggers transition from the S_0 to the S_n states where "n" is greater than 1, which in turn is followed by a relaxation of the molecules to reach the first excited singlet states, before emission takes place. In such cases, the emission spectrum obtained is very much different from the excitation one.

10.4.4 Fluorescence and light diffusion

The energies associated with the diffused and excited photons are equal, which creates problems in obtaining the true spectral layout of the fluorescence effect. The diffused light spectrum generally includes two types of peaks, which are the Raman and the Rayleigh spectra. Since the energies are equal, the Rayleigh line generally superimposes on the excitation line due to both having the same wavelength. In order to avoid this, the excitation wavelength is generally not selected before or at the wavelength of the Rayleigh line.

Though the Rayleigh line can be avoided by this technique, Raman lines are simply unavoidable since these lines exhibit at a wavelength slightly higher than the excitation one. Also, as the diffused light exhibits a lower quantum energy than the excited light, the Raman lines are lower in their intensities than the Rayleigh lines. While measuring the intensity of the fluorescence or excitation spectrum, one always needs to subtract the Raman spectrum from the observed emission spectrum to get the actual fluorescence spectrum. Thus, the higher the intensity of the fluorescence, the intensity of the Raman lines reduces.

For an aqueous solution, Raman peaks are generally caused due to the presence of O-H bonds, the position of which can be determined by the below equation:

$$\frac{1}{\lambda_{ram}} = \frac{1}{\lambda_{ex}} - 0.00034$$

where λ_{ram} is the Raman peak wavelength and λ_{ex} is the excitation wavelength, all in nanometers.

10.4.4.1 Fluorescence lifetime

Molecules, after excitation by a high-energy photon beam, remain in their excited state for a period of time before decaying back to their ground state, resulting in fluorescence. The mean time taken by all the molecules to remain in their excited state is generally termed as the fluorescence lifetime, which ranges from nanoseconds to picoseconds. Fluorescence lifetime is determined, taking into considerations all the radiative and nonradiative processes in the fluorophore de-excitation.

Due to the involvement of a number of processes that simultaneously occur in the fluorescence phenomena, the radiative lifetime, t_r, which is basically the real emission lifetime of a single process that is independent of all other processes (occurring parallel), is hard to be determined. A time characteristic of all de-excitation processes is measured, which is basically the fluorescence lifetime, and is lower than the radiative lifetime. The fractional contribution of each lifetime (f_i) in a fluorophore having several fluorescence lifetimes (τ_i) for calculating the mean fluorescence lifetime (τ_0) can be given as below:

$$\tau_0 = \sum f_i \tau_i \tag{10.1}$$

$$f_i = \frac{\beta_i \tau_i}{\sum \beta_i \tau_i} \tag{10.2}$$

where β_i is the pre-exponential term.

Various factors that determine the number of fluorescence lifetimes in a fluorophore include the ground-state heterogeneity arising out of the presence of equilibrium between conformers having different fluorescence lifetimes, the internal motion of the protein, the different nonrelaxing states, the presence of components in the macromolecule having different emissions, and the structure of the fluorophore itself. The intensity of fluorescence or the fluorophore population in the excited state (N_0) generally decreases exponentially with time (t), which can be expressed as a differential equation given below:

$$-\frac{dN_t}{dt} = (k_r + k')N_t \tag{10.3}$$

where k_r is the radiative rate constant, k' is the sum of all the rate constants of all the processes, and "t" is the time when the population is N_t.

$$\frac{dN_t}{N_t} = -(k_r + k')dt \tag{10.4}$$

At $t = 0$, $N = N_0$, therefore

$$\log \frac{N}{N_0} = -(k_r + k')t \tag{10.5}$$

$$\frac{N}{N_0} = e^{-(k_r + k')t} \tag{10.6}$$

$$N = N_0 e^{-(k_r + k')t} \tag{10.7}$$

Considering

$$\tau = \frac{1}{(k_r + k')}, \text{ we have } N = N_0 e^{\frac{-t}{\tau}} \tag{10.8}$$

when

$$t = \tau, \ N = N_0/e \tag{10.9}$$

Thus, the fluorescence lifetime is basically the time period at which the population of the excited states is reduced by a factor of "e".

10.4.4.2 Fluorescence lifetime measurements

The techniques used in lifetime measurements can be broadly classified into two types: time-based domain measurements and frequency-based domain measurements. Time-based domains are generally able to take into account the time factor and can directly measure the decay whereas frequency domains exhibit the intensity of the spectrum with respect to the wavelength; so examining the decay becomes an implicit case. Techniques like Strobe and Time-correlated single photon counting (TCSPC) comes under the time domain lifetime measurements while multi frequency and cross correlation spectroscopy comes under the frequency domain.

1. Time-correlated single photon counting (TCSPC): Here, a pulsed light source is generally used and the optics and detectors are arranged such that only one photon is detected for a given sample. A timer is switch ON when the light source is started and it is measured when it reaches the detector. Thus, a fluorescence decay curve is generated by counting the number of photons received with respect to the time taken.

2. Strobe or pulse sampling technique: Here, a pulsed light source is used to excite the sample and a narrow time window is opened at each pulse to collect the photons that are eventually saved into the computer. Thus, a decay curve of emission intensity with respect to time can be constructed. The voltage pulse in the photomultiplier tube generally remains synchronized with the pulsed light source.

3. In frequency domain instruments, a phase angle separation between the sinusoidal modulated emission intensity (w) at the detector and the sinusoidal intensity from the light source is generally measured, with both (excited and emitted light) having the same wavelength. The intensity of the emitted light is always lower than that of the excited light and the difference in their phases or the phase angle (denoted as "wt" where "t" represents the delay time of the phase) can be equated

with the fluorescence lifetime of the sample or decay time. A large phase shift is generally observed for a material having a large fluorescence lifetime.

Two identical detectors are generally used to examine the output of the sample fluorescence photomultiplier and the reference photomultiplier, thus measuring the phase shift and the modulations of both the scattered and the fluorescence light. Emission lifetimes can be achieved by either the phase method or the modulation method.

In phase modulation, a high frequency modulation of the excitation beam is achieved. The output of the photomultiplier that detects the fluorescence of the sample records the phase and compares it with the phase of the excitation beam that is separated out as input to the reference photomultipliers. The relation between the phase (p) with respect to the excited beam can be expressed for an emission decaying exponentially with lifetime (τ), as below:

$$\omega\tau = tanp \tag{10.10}$$

where ω is the frequency of modulation. Thus, emission lifetimes can be obtained in this way.

Multifrequency technique is generally achieved by frequency modulations, done using an electro-optic modulator that is crossed transversely by an electric current.

The outputs from both the reference and the detection photomultipliers are passed on to a sinusoidal electric current $R(t)$ having frequency ($\omega + \Delta\omega$) as close as possible to the fluorescent frequency, ω, where $\Delta\omega$ is the cross correlation frequency. The cross correlation frequency can be adjusted accordingly, which possesses all information of the fluorescent light [76–77].

10.5 Examples on fluorescence studies for polymeric nanocomposites

In this section of the chapter, the fluorescent studies associated with polymeric nanocomposites with nanofillers, exhibiting semiconducting characteristics, have been depicted. The color of the emitted fluorescence is tunable in accordance with the geometry of the embedded nanoparticles. For instance, orange and yellow light emissions are observed when the polyvinyl alcohol is embedded with CdS nanoparticles. In another research study, the maximum fluorescence was observed at 630 nm for the polyethylene oxide nanocomposites with the embedded CdS nanoparticles [78]

Few other studies have investigated the effect of the nanoparticle size on the emitted fluorescence. A broad range concerning fluorescence emission has been revealed (red to blue color), depending on the size of the CdS and CdSe/ZnS nanoparticles in the poly methyl methacrylate matrix [79]. Fluorescence at 530 nm has been revealed in the case of CdS nanoparticles within the poly(n-methylol acrylamide) matrix. Fluorescence at 350 nm was attributed to the polymeric matrix [80].

ZnO has also been employed in various instances to produce fluorescence composite materials. Fluorescence of the polymeric nanocomposites with ZnO nanoparticles has been observed within the UV region and as such, these nanocomposites have been employed for a variety of applications as for instance, in corrosion detection system. ZnO-embedded nanocomposites with PMMA, epoxy, and polyurethane have been reported widely. The formation of ZnO-based polyurethane, rich in triazole, has been revealed around 350–400 nm within the UV region (Fig. 10.6). As such, these composite materials have been employed widely for early corrosion detection devices.

The spectra concerning the UV absorption and fluorescence emission has been depicted in Fig. 10.6 for the ZnO-embedded nanoparticles. The $\pi \rightarrow \pi^*$ electronic transitions concerning the pyrene atomic ring corresponds to the 266, 277, 328, and 34 wavelengths. For the ZnO nanoparticles, this peak corresponds to 374 nm. The FTIR spectra revealed the disappearance of the stretching frequency associated with the PyZnO, and this observation, in tandem to that made earlier, is suggestive of the reactions between the molecules of the ZnO nanoparticles synthesized using azide and that of propargyl pyrene. An increase in the intensity associated with the absorption of UV has been revealed with the increasing concentration of the pyrene units on the surface of the ZnO nanoparticles. The highest intense peak from the fluorescence emission data has been revealed at wavelengths of 380 and 415 nm. These high-intensity wavelengths are associated with the emission of the pyrene excimer emission. The lower intense peaks at 480 nm and 510 nm in the fluorescence emission data corresponded to the pyrene excimer emission. This could be attributed to the immobilization of the pyrene units onto the hydrophobic surface of the ZnO nanoparticles. The observations suggest the formation of the ZnO nanoparticles that are fluorescent-active within the UV region. Hence, it can be deduced that to characterize the attachment of fluorophore onto the ZnO nanoparticles, fluorescence spectroscopy can be effectively employed.

The nanocomposites, derived from PyPU in different weight percentages and polyurethane as matrix, have also been investigated for the absorption of the UV radiations and have been characterized using the fluorescence spectroscopy. Figure 10.7 reveals the investigated spectral analysis for the PyPU nanocomposites. An improvement in the absorption intensity of the fabricated films has been revealed for the increased loading of the PyZnO nanoparticles. The peaks around 250–300 nm and 320 nm correspond to the presence of pyrene units, while that for the ZnO nanoparticles, the peak could be observed at 340 nm. Similarly, in the case of fluorescence spectroscopy, the associated emissions have been revealed to increase with the increased loading of the pyrene-grafted ZnO nanoparticles. Studies revealed very good fluorescence emission between the wavelengths 350 nm and 400 nm. Fluorescence emissions of the pyrene monomers have been observed in the case of the polyurethane films. On the other

266, 277, 328, 344 nm
π→π* transitions of pyrene

374 nm
ZnO nanoparticles

380 and 415 nm
the pyrene monomer emission

480 and 510 nm pyrene
excimer emission

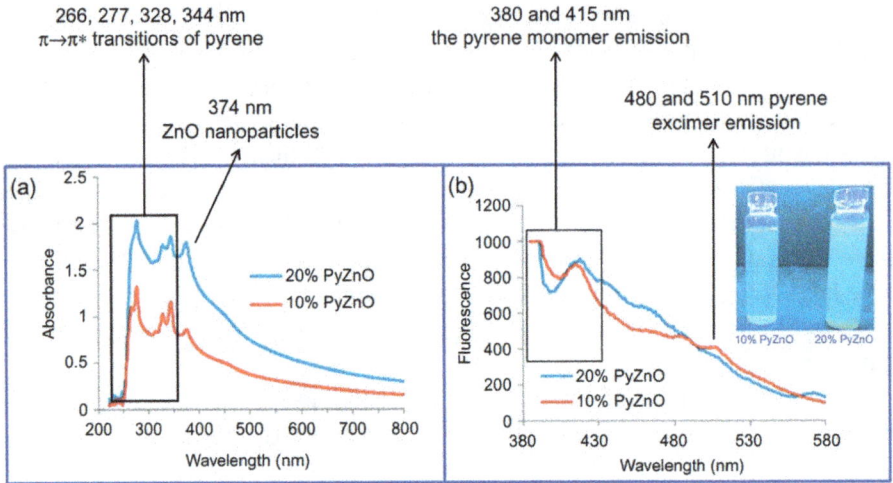

Fig. 10.6: UV spectral and fluorescence spectroscopy of pyrene ZnO nanoparticles (reproduced with permission from [81]).

hand, the fluorescence associated with the pyrene excimer has been revealed to be quenched, and this may be attributed to the presence of urethane groups exhibiting hydrophobic nature.

Around 250–300 and 320 nm
correspond to pyrene units

Peaks observed around 340 nm
correspond to the ZnO nanoparticles 350 and 400 nm
(in near UV region)

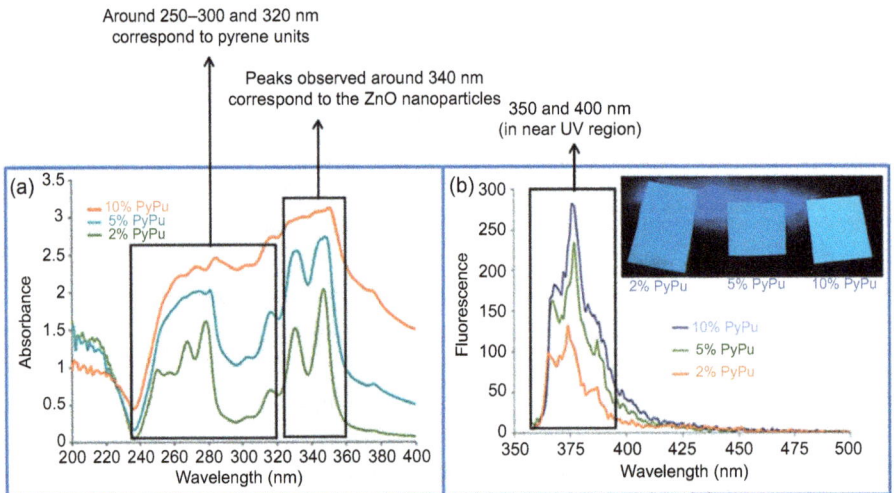

Fig. 10.7: UV spectral and fluorescence spectroscopy of polyurethane films (reproduced with permission from [81]).

10.6 Conclusion

Fluorescence spectroscopy is one of the widely employed characterization techniques amongst the scientific community associated with polymeric nanotechnology. A wide range of information can be revealed from fluorescence spectroscopy, such as the molecular processes, the distance between the nanoparticles, the interaction between the nanoparticles and the polymeric matrix, etc. The use of fluorescence spectroscopy is further enhanced with the recent technological advancements in the domain of cellular imaging, detection for single molecule, and drug delivery. One of the major advantages of employing fluorescence spectroscopy is the reduction in the need of complex technologies and costly instruments. Therefore, it can be concluded that fluorescence spectroscopy will contribute to the rapid advancements of different domains of nanotechnology.

Chapter 11
Nuclear magnetic resonance spectroscopy

11.1 Introduction

Nuclear magnetic resonance (NMR) spectroscopy has been widely used by chemists for structural determination of a number of organic or biological and even inorganic molecules. It has been proven to be the most sophisticated and effective tool after numerous internal developments and changes over time. It is capable of studying molecules not only in their pure state but also in the form of mixtures and in both solids and liquids.

NMR basically studies the behavior of atomic nuclei in their spin state, which induces a magnetic moment upon application of external magnetic field. When a steady and homogeneous field is applied across a sample, the nuclei in the molecules in their spin condition tend to exhibit certain energy levels, which trigger transition between different energy states based on the absorption of suitable radio frequency (rf) radiation. A few of the most suitable nuclei for studying using NMR spectrometer include ^{1}H, ^{19}F, ^{13}C, and ^{31}P.

11.2 Instrumental details and working

Nuclear magnetic resonance (NMR) [82–83] is a spectroscopy technique where the spin of the nucleus in a sample atom is analyzed by placing the sample in a strong magnetic field and excitation by the radio frequency radiation directed perpendicular to the magnetic field vector. NMR spectroscopy is mainly of two types, which are the continuous wave (CW) spectrometer and the Fourier transform spectrometer. In CW spectrometer, the major components included are:
1. A powerful magnet.
2. A probe (consisting of the radio frequency transmitter and receiver).
3. Sweep coils and recorder.
4. The radio frequency transmitter is used to supply Larmor frequency in the x-direction considering the direction to be perpendicular to the applied magnetic field and depends on its strength and nucleus charge. The radio frequency receiver is used to detect resonance signals in the y-direction of the rotating frame by registers. The recorder provides the signals obtained in the receiver.

In order to search for the resonance frequency in a particular spectroscopy, sweeping of the radio Larmor frequency or the applied magnetic induction B_0 is performed, keeping the magnetic induction or the radio frequency constant, respectively. A search for the resonance frequency can also be made by applying a strong pulse of the radio frequency in

https://doi.org/10.1515/9783110997590-011

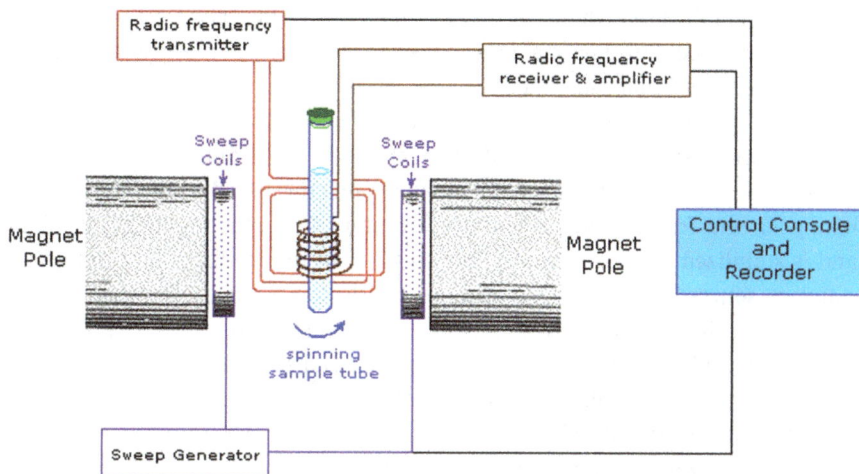

Fig. 11.1: Components of NMR spectrometer. Source: schematic diagram of NMR spectrometer (retrieved on 20 February 2020), Pharmatutor, Pharmacy Infopedia (reproduced with permission from [84]).

a very short duration of time and observing decay in the free induction. In contrast, continuous wave (CW) spectrometers generally use a continuous train of signals.

11.2.1 Powerful magnets

The sample is basically placed in between the powerful magnets as shown in Fig. 11.1 where a net magnetization of the nucleus is induced by the applied magnetic field B_0. Permanent magnets are generally operated at ca 1.4 Tesla that corresponds to 60 MHz of resonance frequency in ^1H and 56.4 MHz in ^{19}F and are designed for "routine"-type spectrometer rather than a research one. In the case of higher magnetic field requirement, permanent magnets are generally not preferred as they suffer from strength limitation. Perkin-Elmer spectrometers were used, which employed field strength of 2.35 T and accordingly corresponds to the 100 MHz of the resonant frequency in ^1H and 94.1 MHz in ^{19}F. In a newer version, Perkin-Elmer R32 can provide field strength of 2.1 T that corresponds to the resonant radio frequency of 90 MHz in ^1H and 84.7 MHz in ^{19}F. This is basically the limit of strength for spectrometers using permanent magnets.

Powerful electromagnets can address the limitations in permanent magnet spectroscopy that exhibits a flexible and a stable field operating in the range of 1–2.5 T. The stability of the field in the best permanent magnet-employed spectroscopy can be realized. These electromagnets come in a variety of field strengths such as 1.4 T, 2.1 T, and 2.35 T found in commercial markets and are used for both routine and research purposes. For very high field strength of about 7 T, cryogenically stored superconducting magnets are generally used. These generally do not require any corrective devices and can maintain high stability of permanent magnets in the field once it gets established.

Two of the most important characteristics of the magnetic field required in an NMR spectroscopy are field stabilization and field homogeneity.

i.   Magnetic field stabilizing is mainly dependent on the stabilized current flow through the coil, which is supplied from the mains through the regulators. Alternating current from the mains is converted into DC current supply by rectifiers and is stabilized using a feedback loop system. In the feedback loop system, a standard source of high stability such as the Weston cell is used to compare its voltage with the voltage across a resistor connected to the D.C. circuit of the coils. The difference in the voltage values is amplified and sent as an error for voltage correction. This is basically the first-order correction to the magnetic field input, exhibiting a field stability of about one part in 10^6.

Flux stabilizers are used to improve the current stability to about one part in 10^8, which is able to provide correction for a short term or rapid fluctuations, though it fails to provide correction for slower and long term fluctuations in field strength. Correction coils are generally wound around the pole pieces of the magnet, which is sensitive to change in flux across the gap between the magnet poles. These changes in flux induce a voltage in the coil, which is amplified and sent as an error signal for correction, resulting in high stability in the current through the main coils and the magnetic field generated.

In order to tackle the slower and the long term fluctuations in field strength, third-order correction methods are generally employed where a reference signal is kept in resonance to the "locked-in" applied magnetic field B_0 by adjusting the radio frequency of the transmitter. Two methods are mainly applied for this. In the first method, a reference sample (H_2O) is kept in the same "locked-in" magnetic field and the radio frequency obtained in the standard probe is compared with the resonance signal obtained in the control probe. An error signal is thus generated when there is a slight drift in the field or the radio frequency, which is then amplified in the main magnet power supply or fed to the flux stabilizer. In the second method, two sample probe systems are discarded with only the sample probe and the sample into which the standard compound is embedded. This technique was first developed by Primas [83]; it is sophisticated and also widely used. The standard compound used in the sample (hexafluorobenzene for ^{19}F sample and tetramethysilane for ^1H are generally used) exhibits a sharp signal, which is far from the resonances obtained in the sample signals and the dispersion error signals obtained are fed to the flux stabilizer. The flux stabilizer then corrects the drift in signals by maintaining a constant value of the field/frequency ratio over a long period of time. The standard compound may be of the same kind as that of the sample or different, depending on the type of spectrometer.

ii. Magnetic field homogeneity consideration in the region of space occupied by the sample is necessary for achieving a high resolution signal. The homogeneity condition desired lies in the range of one part in 10^8 to 10^9, and is achieved by precisely determining the pole piece designs involving precise parallel magnetic field lines, uniformity in metallurgical content, and designs, devoid of any machining marks.

Fine coils such as Shim coils and Golay coils are used on the pole faces in a specific pattern and current is passed, which can be varied using manual controls to correct the overall field homogeneity. Apart from all these corrections, spinning of the sample at a frequency of about 20 Hz about an axis vertical to the horizontally applied magnetic field can further reduce the inhomogeneity by averaging out the effects.

11.2.2 Probe

A probe is basically used to hold the sample using a holder, which also acts as a turbine and ejects out pressurized air tangentially to the sample piece, thereby spinning it about the vertical y-axis. It is available in two types, that is, in crossed-coil and single-coil varieties.

In the crossed-coil variety, two orthogonal axes coils are used with one acting as a transmitter and the other as a receiver. A radio frequency (rf) signal is emitted by the transmitter coil onto the vertically spinning sample along the direction of the magnetic field. By varying the frequency of the emitted rf or the magnetic field strength until resonance is reached, the net magnetization of the nuclei in the sample is obtained, which tends to generate an induced rf signal that is picked up by the rf receiver coil to the recorder.

Single coil probes consist of a single radio frequency (RF) coil. This coil both transmits RF pulses to the sample and detects the resulting NMR signals. Single coil probes are commonly used in routine NMR experiments where simplicity and ease of use are important. Single coil probes are easier to operate and require less tuning and matching and more cost-effective compared to multi-coil designs.

11.2.3 Sweep coils and recorder

In order to achieve resonance condition, the magnetic field and the rf frequency must be varied until resonance is obtained on the recorder spectrum. This is done by either a frequency sweep or slow field. Sweep coils consisting of a pair of Helmholtz coils on either side of the sample with their axis along the magnetic field are used. A saw tooth signal is fed onto these coils, which allows sweeping of the field for resonance in a uniform and a recurrent manner. Sweeping is done to obtain an oscilloscopic output for references of the resonances in different nucleuses of the sample, during probe tuning and field homogeneity optimization. The recorder helps in sweeping the rf and the field as required, by providing the necessary voltage ramp enabling usage of pre-calibrated charts and ensuring correlation between the two.

11.2.4 Few other useful accessories

A few other spectroscopy accessories that are needed to analyze the chemical structure of the compound in complete form or for more information to be derived are discussed below.

i. Spin decoupling: Due to magnetic interactions between the spin -1/2 nuclei in the same molecule a number of resonance lines can be observed in the NMR spectrum. The fine structures on resonance lines basically represent the different energy levels in a spin system, which is also known as spin-spin coupling. In order to decouple the spin system, a secondary rf field is generated, which irradiates the coupled nuclei at the start of the multiplet on the spectrum and results in complete vanishing of the spin-spin coupling lines. The secondary rf source is equipped in most of the spectrometers that can be targeted at a specific position and is often called spin decoupler. In order to precisely irradiate a single line on the spectrum instead of a group of lines, the secondary rf irradiation must be derived from the same basic spectroscopy rf source with the same specification. Spin decoupling can be of two types: homonuclear and heteronuclear. In homonuclear decoupling, decoupling occurs between the nuclei of the same kind or atoms, while in a heteronuclear decoupling, decoupling occurs between the nuclei of different kind or atoms as achieved in the case of decoupling between the ^{13}C and ^{1}H nuclei. Thus NMR spectra are obtained, displaying vacancy in the region of irradiation by the secondary rf.

ii. INDOR spectroscopy: Internuclear double resonance spectroscopy is an important part of a spectrometer, which involves double irradiation experiments for spin-spin and spin decoupling. The objective is to examine the intensity of a specific spectral line irradiated by the rf field B_1 while sweeping through all other frequency resonances with the help of the secondary rf field B_2. This in turn displays the intensity of the specified line w.r.t. the frequencies of the B_2.

Variable temperature accessories are devices which are used to control the temperature over a wide range to be maintained at a fixed value over a long interval of time, accurately.

Fourier transform (FT) NMR spectrometer is a computerized version of the conventional CW spectrometer with all the major components being the same such as the magnet, radio frequency transmitting and receiving coils, the probe, and the recorder. Integration of computers in the system, various interfaces, pulsing unit of the radio frequencies, high power amplifiers, data acquisition, oscilloscopic display, and data handling are a few additional features in an FT spectrometer, which enhances quality and convenience of analyzing the spectrum and controls the timing of the pulses and system. Instructions are generally input to the computer via a teletype and command generated by a master program. Backups are provided in the form of disc units and magnetic tapes for all the unprocessed data, spectrum data, and programs for analysis to be carried out at any time and for future references. The sequence of experimental operation performed in an FT NMR spectrometer is as follows:

i.   First, a high power rf is sent to the transmitter coil from a high power amplifier via a pulsing unit. The frequency is kept close to the Larmor frequency of the nucleus to be studied. All these commands are executed via computer. The computer again sends command after some 10–100 μs to shut down the rf source generating a pulse.

ii.   A time period of about 1–5s as instructed by the computer is taken as acquisition time of the observing receiver for free induction decay (f.i.d.) collection.

iii. The analog signal received via a low pass filter is then converted into a digital one by using analog-to-digital converter and stored in the computer.

iv.   The procedure is repeated and a fresh digital signal spectrum is added to the previous one.

v.   As the signal received gets added after repeated pulse and then stored in a computer in the time domain, Fourier transform analysis is performed to convert it to the frequency domain spectrum.

vi.   The digital spectrum observed is then converted to analog spectrum using a digital-to-analog converter and the output is recorded on a chart with the intensities and positions being listed corresponding to the digital output.

11.3 Sample preparation, loading, and handling

11.3.1 Sample preparation

In NMR spectroscopy, samples are presented in liquid or solution form in inert solvents while solid samples are hard to examine due to broadening of spectral lines. Gases may be examined but under a specified pressure as given for standard sampling. Samples should be free from any suspended particles or impurities, which are generally responsible for broadening of the spectral lines but can also prevent saturation. For further enhancement in the quality of the spectral lines, the solution sample can also be made free of dissolved oxygen.

For liquid samples, glass cylinders are usually employed to carry the solution, which are fixed onto a turbine arrangement and mounted onto the probe. The turbine makes use of pressurized air to spin the sample within hollow tubes that are precisely bored to have an outer diameter of 5, 8, 10, and 12 mm. Some of the important operations that need to be carried out before carrying out NMR spectroscopy can be described as below.

11.3.2 Sample tube placement

One of the major requirements in an NMR spectroscopy is the sample tube placement, which is critical in a number of ways. NMR tube placement is mainly carried out in two ways. In most of the cases, a depth chart containing the diagram of the spinner turbine and an NMR tube is generally provided. This is referred while inserting the sample tube against the depth chart up to a depth that matches the bottom with the diagram. The other way is to use a depth gauge generally available with spectrometer manufacturers into which the sample tube is inserted until it reaches the bottom of the depth gauge.

Proper place is necessary because if the sample tube is not inserted to the required depth, some of the sample would remain outside the visibility range of the receiver coil, while on the other hand if it is inserted beyond the required depth the expensive probe could itself get damaged with some sample out of the range of the receiver as well. A few other major devices such as temperature sensor, glass inserts, etc. are also equipped into the probe at a depth below the assigned sample depth.

11.3.3 Referencing

All frequencies of lines in NMR spectroscopy are referred based on an arbitrarily selected standard spectral line. The standard resonance spectral line should have sharp, distinct features and at a frequency far away from the typical sample resonance lines like in the case of analyzing the sample for ^{1}H, tetramethylsilane, and CS_2 is convenient where ^{13}C absorption is used as a reference in the study of ^{13}C.

The reference compound can be used along with the sample in the probe by two different ways: internal standard and external standard. In the internal standard arrangement, the reference is basically dissolved into the sample that is under investigation while in an external standard arrangement, a separate capillary for the reference compound is used, which is then inserted being centrally placed into the main sample glass tube. However, in the second situation, the frequency differences need to be adjusted based on the different magnetic susceptibility of the reference and the sample. The frequency difference can be measured directly based on the pre-calibrated charts of a modern spectrometer or by measuring directly the frequency difference using high-precision frequency counter.

11.3.4 Integration

In a general spectroscopy spectrum, the area under the absorption line is basically dependent on the number of nuclei participating in the energy absorption. The area can be calculated by integrating the trace line for a given interval of frequencies.

NMR spectroscopy is generally arranged to give output in the form of the integral or area under the absorption peak.

11.3.5 Field/frequency locking

Field/frequency locking is generally established by employing a separate compound (apart from the sample to be determined) in obtaining its resonance frequency so that drifts in magnetic field can be avoided. As the nucleus to which the frequency is to be locked cannot be of the same type as the sample nucleus, ^{1}H cannot be employed for this purpose. Deuterium, however, can serve the purpose of field/frequency locking as it can easily replace protons from most of the organic compounds and is widely available commercially in a variety of deuterated solvents. This type of locking is generally termed as "internal lock." Here, the deuterated solvent sample is irradiated using lock transmitter coils perpendicular to the magnetic field and the frequency is varied for obtaining the locking frequency corresponding to the resonance frequency of the deuterium nuclei. The observer notices a strong dip in the interference pattern sine waves at the resonance or locking frequency due to cancelling out with the deuterium frequency and then turns ON the lock. In most of the modern NMR spectrometers, this operation is automated and the device itself searches for the frequency of resonance and lock it.

The lock signal stability is basically dependent on certain factors, that is, the power of the lock transmitter, the homogeneity in the magnetic field, and the phase of the lock signal. When the lock signal is observed to be erratically fluctuating up and down it is interpreted as presence of impurities in the sample. Conversely, a regular oscillation in the lock signal can be understood as indicating that the lock signal has reached a saturation state, often caused by excessive power. The solution to this issue is to reduce the power. During saturation, generally the lock signal tends to initially drop and then rise steeply. As the instabilities in the lock signal cannot be avoided completely due to presence of suspended particles and saturation conditions, they generally interfere with the magnetic field and might even result in loss of lock signal. After a stable range of lock signal is reached, the lock phase needs to be maximized as it depends on the homogeneity of the magnetic field.

11.3.6 Spectrometer operation

In a CW spectrometer, the field/frequency lock circuit generally places the resonance frequency line of the reference toward the right hand side of the recorder chart. The frequency of the sample is then varied as decided by the operator along with the power required in the observing field. The obtained spectrum is then analyzed and made free of noise; the spectrum size is increased by amplifying the signal and suitable phase setting is chosen. From the extreme left hand side of the recorder, either the field is swept from low to high field or the frequency swept from high to low frequency.

11.3.7 Measurements of relaxation times

Measurements of the relaxation times are necessary in case of a pulse spectrometer. The aim is to set the computers to perform the NMR experiment for a suitable multi-pulse sequence. For a spin lattice relaxation time (T_1) measurement, a pulse is generated, which inverts the stable state of the magnetization in the z-axis (toward the applied magnetic field) by 180° toward −z-axis. While returning again to its stable Boltzmann equilibrium state, a 90° pulse is generated after time "t," which aligns the magnetization vector along the y-axis. The induced signal is recorded, which measures the T_1 relaxation time representing the decay by spin lattice interaction. The technique can also be termed as 180°-t-90° pulse sequence. Considerable time has to be given between sequences while enhancing the signal using repeated sequence and is generally of the order of five times the T_1. When the intensities of different lines are plotted with sequential values of t, a plot showing an experimental increase w.r.t. to time constant T_1 is observed.

For T_2 measurement, a more complex pulse sequence such as the spin echo sequence of Hahn, which can be represented as 90°t-180°-2t, modified by Carr Purcell and further developed by Meiboom Gill was required.

11.4 Analysis of NMR spectra

Analysis of NMR spectra is not limited to mere measurements in position and intensities but also used to determine few fundamental parameters, that is, chemical shifts and coupling constants.

11.4.1 Chemical shift

The Larmor frequency of the nucleus is dependent on its electronic environment (chemical properties), that is, the role of the atom in which the nucleus is present. For a stable magnetic field, different compounds generally exhibit separate frequencies of absorption signals as can be seen in the case of water, benzene, and cyclohexane where single absorption lines are exhibited at particular frequencies. This phenomenon is generally referred to as chemical shifts (Fig. 11.2). Chemical shifts can also vary for nuclei with different positions in the same molecule such as in ethanol where ^1H nuclei exhibits three different absorption signals having relative intensities of 1:2:3

for OH, CH_2, and CH_3, respectively. Chemical shifts are dependent on the field strength and are proportional to it (B_0). The net field (B_{eff}) observed by the nucleus from the magnetic field that is usually screened by the electrons within the molecules can be given as:

$$B_{eff} = B_0(1-\sigma) \tag{11.1}$$

where σ is the screening constant and the difference in the screening constant of two nuclei is generally termed as the chemical shift between the two. This is measured in units of frequency (Hz) irrespective of determination by frequency or field sweep. Chemical shift can also be expressed in terms of parts per million (p.p.m.) given as:

$$\text{chemical shift (p.p.m.)} = \frac{\text{chemical shift (Hz)} \times 10^6}{\text{observation frequency (Hz)}} \tag{11.2}$$

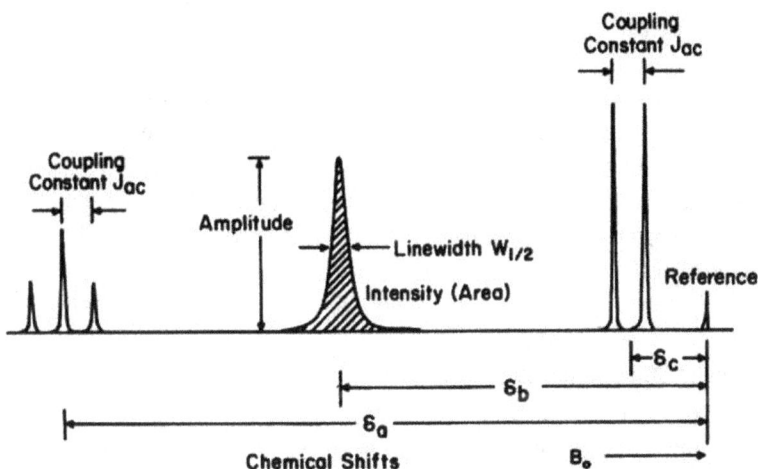

Fig. 11.2: Chemical shifts and coupling constants Source:(reproduced with permission from [85]).

11.4.2 Coupling constant

Chemically shifted lines are observed to be exhibiting multiplet structure, which is due to the interaction between the nuclei in the molecule via the electrons in the orbital of the atom, also generally referred to as spin-spin coupling. A fine splitting in the spectrum of both the coupled nuclei is generally observed, which is dependent on the spin quantum number I. The two coupled-1/2 spin nuclei are seen to be exhibiting doublet when chemically shifted from each other, the distance between which remains constant and are measured in units of frequency (Hz). This separation is independent of the field strength and is referred to as spin-spin coupling constant, J.

No mutual splitting of lines due to spin-spin coupling is observed when equivalent nuclei (as in the case of methyl group) are chemically shifted. However, multiplet structures can be observed for the nuclei interacting with the external groups or molecules, the separation of which is the coupling constant, J. The number of splitting lines observed when external groups are involved in the formation of multiplet structure is generally $(n + 1)$, where "n" is the number of nuclei.

These chemical shifts and coupling constants can be easily recognized directly from the spectrum as long as the separations between the chemically shifted lines are sufficiently large as compared to the coupling constants by a factor of 10 or more. But when the chemically shifted line separation is lower than the factor 10, it becomes hard to identify from the spectrum and different phenomena comes into play such as distortion of the intensities, disappearance, and appearances of expected and unexpected lines. Therefore, a separate spectral analysis is required for such situations.

11.4.3 First-order spectra

In first-order spectra, two nonequivalent coupled-1/2 spin nuclei exhibits double splits of the same magnitude, resulting in four lines. These nuclei can also independently be coupled to other nearby nuclei producing extra splitting lines. Therefore, three groups of four lines generally occur in an AMX system or the first-order 3-spin system. The separation between the centers of the multiplets thus can be denoted by δ, which represents the chemical shift while the separation within the multiplets represents the coupling constant, J. For a nonequivalent system, the first-order spectra analysis of the spin nuclei can thus be carried out.

For a chemically and magnetically equivalent system, where the coupled nuclei spins are also independently coupled to other groups or external nuclei, the spectra obtained are not similar as seen in the previous case. A pattern of $(n + 1)$ lines can be observed where the resonance of a single nucleus or a group of nuclei is split by another group (or nuclei) causing a splitting pattern with "n" denoting the number of equivalent nuclei. However, the intensities of the lines in this case are not of the same magnitude and can be represented as the ratio of the binomial coefficient.

To better understand this phenomenon we need to consider the spin states of a group, say A_3, and how these states can influence the spin states of another group with two nuclei, say X_2, which are chemically and magnetically equivalent. First, let us find out the number of ways in which the spin states of A_3 combine within its nuclei, which is in $2^3 = 8$ ways with α and β states allowed in each nucleus. The energy of these eight combinations of spin states in the group A_3 interacts with the spin states (coupling) in X_2 based on the m_z values of each combination. As the nuclei in X_2 are chemically and magnetically equivalent, the spectral lines of both the nuclei superimpose on each other due to which only one nuclei in X_2 can be considered for the coupling between its α and β spin states and the eight energy states of the group A_3. It must be noted that the intensity obtained in such conditions will be twice the intensity obtained while interacting with the single nucleus in X_2. Thus while interacting, the resultant spectrum generally obtained are four lines of intensities 1:3:3:1.

The frequency of transitions can be given as the difference in the energy levels of the eight combinations of nuclei spins, though only few transitions are allowed between the energy levels, which depend on factors given as:

i. Transitions are allowed only when the m_z values of two participating combinations of nuclei spin differ by ±1 where +1 and −1 correspond to absorption and stimulated emission, respectively.
ii. Transitions are allowed when it involves the change of orientation in the spin of only one nucleus and not more than that.

11.4.4 Second-order spectra

As already mentioned above that when the separation between two chemically shifted lines becomes less than 10 times the coupling constant, impurity lines begin to form, expected lines begin to disappear on the spectrum, and distortion in the intensities of the spectral lines are observed. In such situations, second-order spectra analysis is generally preferred and adopted.

If we consider a dual nuclei system exhibiting four lines in an NMR spectrum, based on the second-order spectra the ratio of the intensities of the inner lines to the outer lines forms a characteristic feature. If the lines are drawn from the left in the order of 1, 2, 3, and 4, the ratio can be represented as:

$$\frac{\text{intensity of 2 (or 3)}}{\text{intensity of 1 (or 4)}} = \frac{(1-4)}{(2-3)} \qquad (11.3)$$

The chemical shift δ and the coupling constant J between the two nuclei can be given as:

$$\delta = [(1+4)(2+3)]^{1/2} \qquad (11.4)$$

$$J = (1-2) = (3-4) \qquad (11.5)$$

11.5 Example for NMR of polymeric nanocomposites

HR solid-state NMR has been employed to characterize the starch-clay nanocomposites. Complete exfoliation of the clay particles was observed in case of nanocomposites comprising 2.5 wt% nanoclay concentrations. The exfoliation has been observed for an average interlayer distance of around 40 nm. The $^1HT_1$ values associated with the starch matrix and the presence of moisture content within the sample has been revealed in Fig. 11.3.

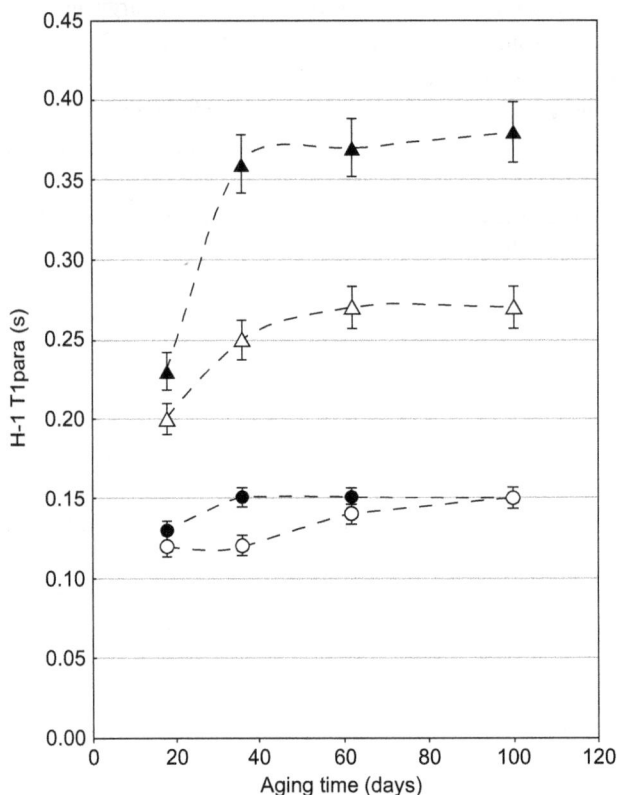

Fig. 11.3: ^{1}HT_1 para data (reproduced with permission from [86]).

An increase in T_1 para values has been revealed for the first 40 days of the prepared sample. This signifies a decrease in the degree of the exfoliation. However, with the passage of the days, that is, from 40 to 60 days, the process became consistent. Overall, it could be deduced that there was no instance seen of the clay dispersion even due to the recrystallization effect of the starch during its ageing process. An increase in T_1 para value associated with the starch matrix has been observed to enhance by 60% during the initial days of the sample ageing with the corresponding interlayer spacing ranging from 33 to 55 nm. The enhanced T_1 para values are significant of the active involvement of the water molecules in retaining a strong interaction with the clay surface.

Figure 11.4 depicts the behavior of the calcium silicate hydrate-based polyvinyl alcohol nanocomposites that have been characterized using the ^{1}C CP NMR methodology. The methane carbon resonances are observed at different peaks, that is,

77 (peak I), 71 (peak II), and 65 ppm (peak III). This has been attributed to the presence of heterotacitic triads within the two intramolecular hydrogen bonds. The methyl signals have been seen at 46 ppm in the case of the C-S-HPN composite specimen. The two methine carbon resonances are observed at 72 and 66 ppm for the same composite specimen. This has been observed in comparison to the pure polyvinyl alcohol and the degree of tacticity and the gelatin procedure are the determining factors for the interactions between PVA and CSH.

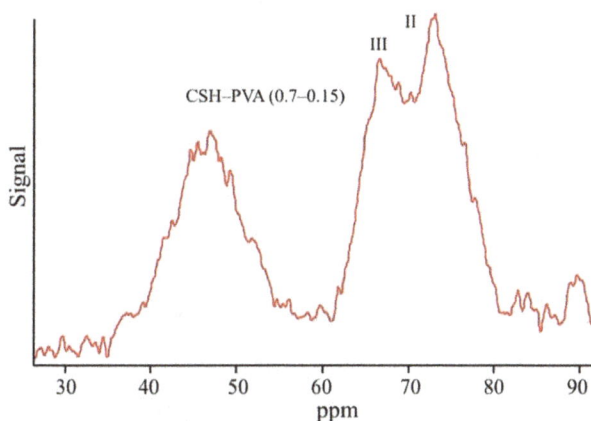

Fig. 11.4: C CP NMR spectra of C-S-HPN material (reproduced with permission from [87]).

^{1}H NMR was employed to carry out characterization of the PLLA/ZnO nanocomposites with varying weight percentages of ZnO concentration. The employability of ^{1}H NMR in the aforementioned case is another instance that corroborates the importance of such techniques in investigating the molecular interactions. The spectrum has been depicted in Fig. 11.5, and it has been observed that the ZnO nanoparticles were distributed uniformly in the PLLA matrix. The spectrum in Fig. 11.5 was obtained by dispersing the nanocomposites in $CDCl_3$. Signals in the spectra were revealed at 1.56 and 5.18 ppm that corresponded, respectively, with the methyl and methane protons in the PLLA.

The methane proton that has been attached to the terminal –OH can be reflected in the weak quartet at 4.35 ppm. The methyl group in the vicinity of the terminal –OH has been observed to be situated at the corresponding 1.35 ppm. The broad peaks ranging from 3.5 to 3.8 ppm are because of the presence of –OH end groups. The optical purity of the PLLA matrix is reflected at the quartet corresponding to 5.18 ppm. This also signifies that racemization was absent during polymerization. The signals at 69.98 ppm and 16.59 ppm in the ^{13}C NMR (Fig. 11.6) also suggested the agreement with the ^{1}H NMR results. These signals were due to the methyl and methane protons. On the other hand, the weak peaks at 66.6, 174.5, and 20.3 ppm corresponded to the methane, carbonyl carbon, and methyl, respectively.

Another peak at 169.57 ppm corresponded to the carbonyl carbon due to the occurrence of highly isotactic polymerization process. The bonding between the carboxylate group and ZnO surface was confirmed from the absence of carboxyl end groups in the obtained ^{1}H NMR and ^{13}C NMR spectra.

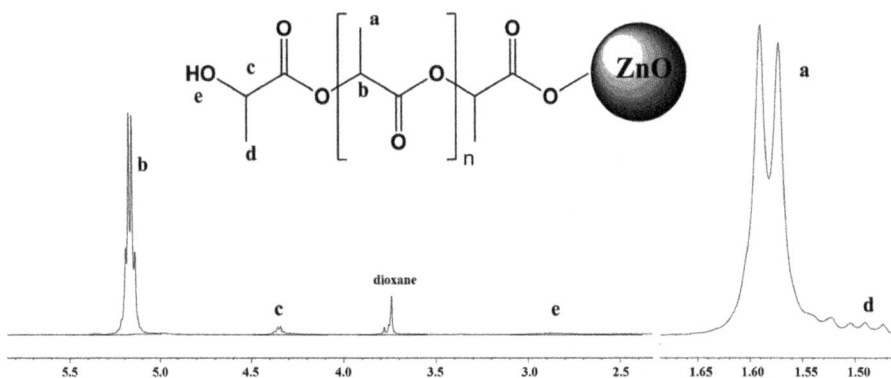

Fig. 11.5: ^{1}H NMR spectrum of PLLA/ZnO nanocomposites (reproduced with permission from [88]).

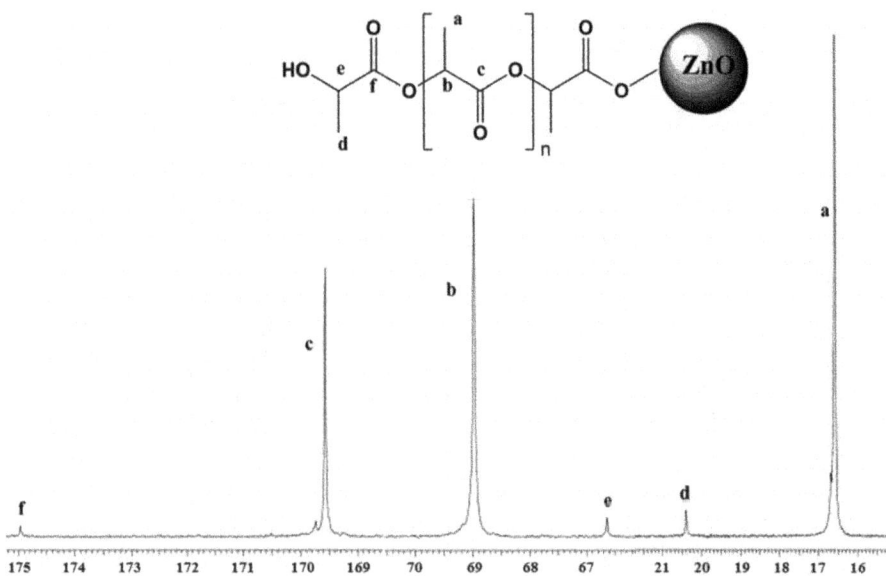

Fig. 11.6: ^{13}C NMR spectrum of PLLA/ZnO nanocomposites (reproduced with permission from [88]).

11.6 Conclusion

The present chapter has discussed the nuclear magnetic spectroscopic technique. The instrument details and the associated operating principles have been discussed to aid practitioners in handling the sophisticated instruments. The example of NMR spectroscopy of the biopolymer nanocomposites reveals that the nature of the embedded filler does not affect the characterization process. However, the NMR spectroscopy is one of the important spectroscopic techniques that aids in understanding different aspects of the polymer-filler characteristics such as bonding, the presence of voids, nature of dispersion, etc.

References

[1]   Sorby, H.C. (1826–1908) Navigational Aids for the History of Science and Technology,https://www. nahste.ac.uk/isaar/GB_0237_NAHSTE_P0261.html (accessed August 2022).

[2]   Corradini, M.G., McClements, D.J. (2019) Microscopy Food Applications, Editor(s): Paul Worsfold, Colin Poole, Alan Townshend, Manuel Miró, Encyclopedia of Analytical Science (3rd edition), Academic Press, Oxford, pp. 47–56, https://doi.org/10.1016/B978-0-12-409547-2.14314-8.

[3]   Brandon, D., Kaplan, W.D. (2008) Microstructural Characterization of Materials (2nd edition), John Wiley & Sons, Ltd, West Sussex, England, https://doi.org/10.1002/9780470727133.

[4]   Bonnamy, S., Oberlin, A. (2016) Chapter 4 – Transmission Electron Microscopy, Editor(s): Michio Inagaki, Feiyu Kang, Materials Science and Engineering of Carbon, Butterworth-Heinemann, pp. 45–70, https://doi.org/10.1016/B978-0-12-805256-3.00004-0.

[5]   Chung, K.T., Reisner, J.H., Campbell, E.R. (1983) Charging phenomena in the scanning electron microscopy of conductor-insulator composites: A tool for composite structural analysis, Journal of Applied Physics, 54 (11), https://doi.org/10.1063/1.331946.

[6]   Zhao, M., Ming, B., Kim, J.-W., Gibbons, L.J., Gu, X., Nguyen, T., Park, C., Lillehei, P.T., Villarrubia, J.S., Vladár, A.E., Alexander, L.J. (2015) New insights into subsurface imaging of carbon nanotubes in polymer composites via scanning electron microscopy, Nanotechnology, 26, https://doi.org/10.1088/ 0957-4484/26/8/085703.

[7]   Vladár, A.E., (2015) Strategies for scanning electron microscopy sample preparation and characterization of multiwall carbon nanotube polymer composites, NIST Special Publication 1200–1217, http://dx.doi.org/10.6028/NIST.SP.1200

[8]   Derjaguin, B., Landau, L.D. (1941) Theory of the stability of strongly charged lyophobic sols and of the adhesion of strongly charged particles in solutions of electrolytes, Acta Physicochimica U.R.S.S., 14, 633–662.

[9]   Verwey, E.J.W., Overbeek, J., Th., G. (1948) Theory of the Stability of Lyophobic Colloids: The Interaction of Sol Particles Having an Electric Double Layer, Elsevier, Amsterdam.

[10] Musumeci, C. (2017) Advanced scanning probe microscopy of graphene and other 2D materials, Crystals, 7 (7), 216, https://doi.org/10.3390/cryst7070216.

[11] Lamas, D.G., (1999) Estudio de las interfaces óxido de conducción iónica/óxido semiconductor/gas— Obtención de un nuevo sensor de gases, Tesis de Doctorado en Ciencias. Físicas, Facultad de Ciencias Exactas y Naturales, Universidad Nacional de Buenos Aires, Buenos Aires, Argentina.

[12] Lamas, D.G., de Oliveira Neto, M., Kellermann, G., Craievich, A.F. (2017) Chapter 5 – X-Ray Diffraction and Scattering by Nanomaterials, Editor(s): Alessandra L. Da Róz, Marystela Ferreira, Fabio de Lima Leite, Osvaldo N. Oliveira, Micro and Nano Technologies, Nanocharacterization Techniques, William Andrew Publishing, Oxford, Pages 111–182, https://doi.org/10.1016/B978-0-323-49778-7.00005-9.

[13] Juárez, R.E., Lamas, D.G., Lascalea, G.E., Walsöe de Reca, N.E. (2000) Synthesis of nanocrystalline zirconia powders for TZP ceramics by a nitrate–citrate combustion route, Journal of the European Ceramic Society, 20 (2), 133–138, https://doi.org/10.1016/S0955-2219(99)00146-6.

[14] Smith, B.C. (2011) Fundamentals of Fourier transform Infrared spectroscopy 2nd edition, CRC Press, Taylor and Francis Group, USA.

[15] Christy, A.A., Ozaki, Y., Gregoriou, V.G. (2001) Modern Fourier Transform Infrared Spectroscopy (Comprehensive Analytical Chemistry), Elsevier Science, London, 35.

[16] https://cdn.shopify.com/s/files/1/0213/2688/products/fasttech-nichrome22_2048x.jpg?v= 1546995706

[17] https://lightingandceilingfans.com/wp-content/uploads/imgp/mercury-arc-lamp-1-4972.jpg

[18] https://eureka.physics.utoronto.ca/Eureka2016/Photos/March2/globarsourceofparis.jpg

[19] https://edisontechcenter.org/NernstLamps.html

https://doi.org/10.1515/9783110997590-012

[20]   https://upload.wikimedia.org/wikipedia/commons/thumb/5/59/Golay_Cell_Schematic.svg/800px-Golay_Cell_Schematic.svg.png

[21]   https://spie.org/Images/Graphics/Publications/FG08_P34_fig2.jpg

[22]   http://www.sciencetech-inc.com/mct-10-010-e-ln-detector-mct-cryogenic-receiver-ln2-cooled-1x1mm.html

[23]   https://www.renishaw.com/media/img/en/e4feffedc4464f74b2dd08dbe14ee89a.jpg

[24]   Jaleh, B., Fakhri, P. (2016) Infrared and Fourier Transform Infrared Spectroscopy for Nanofillers and their Nanocomposites, Editor(s): Sabu Thomas, Didier Rouxel, Deepalekshmi Ponnamma, Spectroscopy of Polymer Nanocomposites, pp. 112–129, William Andrew Publishing, Oxford.

[25]   Nasrollahzadeh, M., Jaleh, B., Jabbari, A. (2014) Synthesis, characterization and catalytic activity of graphene oxide/ZnO nanocomposites, RSC Advances, 4 (69), 36713–36720.

[26]   Fakhri, P., Nasrollahzadeh, M., Jaleh, B. (2014) Graphene oxide supported Au nanoparticles as an efficient catalyst for reduction of nitro compounds and Suzuki–Miyaura coupling in water, RSC Advances, 4 (89), 48691–48697.

[27]   Abedi, K., Ghorbani-Shahna, F., Jaleh, B., Bahrami, A., Yarahmadi, R., Haddadi, R., Gandomi, M. (2015) Decomposition of chlorinated volatile organic compounds (CVOCs) using NTP coupled with TiO2/GAC, ZnO/GAC, and TiO2–ZnO/GAC in a plasma-assisted catalysis system, Journal of Electrostatics, 73, 80–88.

[28]   Jaleh, B., Fakhri, P. (2016) Infrared and Fourier Transform Infrared Spectroscopy for Nanofillers and their Nanocomposites, Spectroscopy of Polymer Nanocomposites, pp. 112–129, William Andrew Publishing.

[29]   Jaleh, B., Shahbazi, N. (2014) Surface properties of UV irradiated PC–TiO2 nanocomposite film, Applied Surface Science, 313, 251–258.

[30]   Ham, N.S., Walsh, A. (1958) Microwave-powered Raman sources, Spectrochimica Acta, 12, 88.

[31]   Stammreich, H. (1956) Technique and results of excitation of Raman spectra in the red and near infra-red region, Spectrochimica Acta, 8, 41.

[32]   Stammreich, H. (1950) The Raman Spectrum of Bromine, Physical Review, 78, 79.

[33]   Stammreich, H., Forneris, R., Sone, K. (1955) Raman Spectrum of Sulfur Dichloride, Chemical Physics, 23, 1972.

[34]   Stammreich, H., Forneris, R., Tavares, Y. (1956) Raman Spectrum of Thionyl Bromide, Chemical Physics, 25, 580, 1277 and 1278.

[35]   Stammreich, H. (1950) Das Raman-Spektrum des Azobenzols, Experientia, 6, 224.

[36]   Rank, D.H., Wiegand, R.V. (1942) A Photoelectric Raman Spectrograph for Quantitative Analysis, Optical Society of America, 32, 190.

[37]   Ferraro, J.R., Nakamoto, K., Brown, C.W. (2003) Introductory Raman Spectroscopy, (2) Elsevier Science, USA.

[38]   https://commons.wikimedia.org/wiki/File:Setup_Raman_Spectroscopy_adapted_from_Thomas_Schmid_and_Petra_Dariz_in_Heritage_2(2)_(2019)_1662-1683.png

[39]   https://commons.wikimedia.org/wiki/File:Raman_scattering.svg

[40]   https://en.wikipedia.org/wiki/Scintillation_counter#/media/File:PhotoMultiplierTubeAndScintillator.svg

[41]   https://commons.wikimedia.org/wiki/File:Photodiode_array_chip.jpg

[42]   https://commons.wikimedia.org/wiki/File:Diagram_of_a_charge-coupled_device.jpg

[43]   Smith, E., Dent, G. (2005) Modern Raman Spectroscopy- A Practical Approach, (2) John Wiley and Sons, Ltd, England.

[44]   Yan, X., Itoh, T., Kitahama, Y., Suzuki, T., Sato, H., Miyake, T., Ozaki, Y. (2012) A Raman spectroscopy study on single-wall carbon nanotube/polystyrene nanocomposites: Mechanical compression transferred from the polymer to single-wall carbon nanotubes, The Journal of Physical Chemistry C, 116 (33), 17897–17903.

[45] Loa, I. (2003) Raman spectroscopy on carbon nanotubes at high pressure, Journal of Raman Spectroscopy, 34 (7-8), 611–627.

[46] Srivastava, I., Mehta, R.J., Yu, Z.Z., Schadler, L., Koratkar, N. (2011) Raman study of interfacial load transfer in graphene nanocomposites, Applied Physics Letters, 98 (6), 063102.

[47] Heide, P.V.D. (2012) X-Ray Photoelectron Spectroscopy: An Introduction to Principles and Practices 1st edition, John Wiley and Sons, New Jersey, 1.

[48] Hofmann, S. (2013) Auger and x-ray Photoelectron Spectroscopy in Materials Science 1st edition, Springer series in Surface science 49, Springer-Verlag, Berlin.

[49] https://www.prevac.eu/Media/catalog/product/6/big_12.jpg

[50] https://cdn11.bigcommerce.com/s-ap31zi/images/stencil/608x608/products/104/204/2__32354.1352963587.JPG?c=2

[51] https://doi.org/10.1007/978-0-387-92897-5 https://media.springernature.com/original/springer-static/image/prt%3A978-0-387-92897-5%2F1/MediaObjects/978-0-387-92897-5_1_Part_Fig3-1224_HTML.jpg

[52] https://www.jobilize.com/ocw/mirror/col10699/m43546/Picture_1.png

[53] https://www.researchgate.net/profile/Andrea_Bazzano/publication/312494038/figure/fig7/AS:669573225402391@1536650105940/discrete-dynode-secondary-electron-multiplier.png

[54] https://www.photonis.com/sites/default/files/inline-images/How-it-works.jpg

[55] https://www.rug.nl/research/portal/files/54830051/Chapter_2.pdf

[56] Mane, A.T., Patil, V.B. (2016) X-ray Photoelectron Spectroscopy of Nanofillers and their Polymer Nanocomposites, Editor(s): Sabu Thomas, Didier Rouxel, Deepalekshmi Ponnamma, Spectroscopy of Polymer Nanocomposites, pp. 452–467, William Andrew Publishing, Oxford.

[57] Turner, D.W., Baker, C., Baker, A.D., Bundle, C.R. (1970) Photoelectron Spectroscopy, Wiley-Interscience, New York.

[58] Svehla, G. (1979) A Comprehensive Analytical Chemistry: Ultraviolet Photoelectron and Photoion Spectroscopy, Auger Electron Spectroscopy, Plasma Excitation in Spectrochemical Analysis, Elsevier scientific publishing company, Amsterdam, 9.

[59] Straughan, B.P., Walker, S. (1976) Spectroscopy, Chapman and Hall Ltd., London, 3.

[60] https://www.semanticscholar.org/paper/A-capillary-discharge-tube-for-the-production-of-Schoenhense-Heinzmann/e4383efe9e6963a4c258536817ad5e9e5a135b35

[61] https://www.researchgate.net/figure/Schematic-representation-of-the-synchrotron-radiation-source-The-electron-storage-ring_fig13_307606559

[62] https://d3pcsg2wjq9izr.cloudfront.net/files/85908/images/235-seya-namioka-monochromator-400.jpg

[63] http://es1.ph.man.ac.uk/GCK2/HDA.html

[64] Zhu, J.F., Zhu, Y.J. (2006) Microwave-assisted one-step synthesis of polyacrylamide– metal (M= Ag, Pt, Cu) nanocomposites in ethylene glycol, The Journal of Physical Chemistry B, 110 (17), 8593–8597.

[65] Chen, L., Zhang, Y., Zhu, P., Zhou, F., Zeng, W., Lu, D.D., Sun, R., Wong, C. (2015) Copper salts mediated morphological transformation of Cu 2 O from cubes to hierarchical flower-like or microspheres and their supercapacitors performances, Scientific Reports, 5, 9672.

[66] Cao, Z., Fu, H., Kang, L., Huang, L., Zhai, T., Ma, Y., Yao, J. (2008) Rapid room-temperature synthesis of silver nanoplates with tunable in-plane surface plasmon resonance from visible to near-IR, Journal of Materials Chemistry, 18 (23), 2673–2678.

[67] Saute, B., Premasiri, R., Ziegler, L., Narayanan, R. (2012) Gold nanorods as surface enhanced Raman spectroscopy substrates for sensitive and selective detection of ultra-low levels of dithiocarbamate pesticides, Analyst, 137 (21), 5082–5087.

[68] Khanna, P.K., Singh, N., Charan, S., Subbarao, V.V.V.S., Gokhale, R., Mulik, U.P. (2005) Synthesis and characterization of Ag/PVA nanocomposite by chemical reduction method, Materials Chemistry and Physics, 93 (1), 117–121.

[69] Chitte, H.K., Bhat, N.V., Karmakar, N.S., Kothari, D.C., Shinde, G.N. (2012) Synthesis and characterization of polymeric composites embeded with silver nanoparticles, World Journal of Nano Science and Engineering, 2 (1), 19–24.

[70] Biswas, A., Aktas, O.C., Schürmann, U., Saeed, U., Zaporojtchenko, V., Faupel, F., Strunskus, T. (2004 Apr 5) Tunable multiple plasmon resonance wavelengths response from multicomponent polymer-metal nanocomposite systems, Applied Physics Letters, 84 (14), 2655–2657.

[71] Villalva, D.G., (2015) Development of new liposome based sensors. DOI: 10.13140/ RG.2.2.26454.55366, https://www.researchgate.net/figure/Schematic-illustration-of-spectrofluorimeter_fig12_319068343

[72] https://www.electrical4u.com/images/september16/1479912879.png

[73] http://elchem.kaist.ac.kr/vt/chem-ed/optics/detector/graphics/pmt.gif

[74] https://photonengr.com/wp-content/uploads/2017/01/wollaston-1024x651.png

[75] https://upload.wikimedia.org/wikipedia/commons/0/00/2_sizes_of_cuvette.jpg

[76] Albani, J.R. (2007) Principles and Applications of Fluorescence Spectroscopy, Blackwell publishing company, UK.

[77] Straughan, B.P., Walker, S. (1976) Spectroscopy, Chapman and Hall Ltd, London.

[78] Khanna, P.K., Gokhale, R.R., Subbarao, V.V.V.S., Singh, N., Jun, K.W., Das, B.K. (2005) Synthesis and optical properties of CdS/PVA nanocomposites, Materials Chemistry and Physics, 94 (2–3), 454–459.

[79] Tamborra, M., Striccoli, M., Curri, M.L., Agostiano, A. (2008) Hybrid nanocomposites based on luminescent colloidal nanocrystals in poly (methyl methacrylate): Spectroscopical and morphological studies, Journal of Nanoscience and Nanotechnology, 8 (2), 628–634.

[80] Fang, Y., Chen, L., Wang, C.F., Chen, S. (2010) Facile synthesis of fluorescent quantum dot-polymer nanocomposites via frontal polymerization, Journal of Polymer Science Part A: Polymer Chemistry, 48 (10), 2170–2177.

[81] Gaddam, R.R., Narayan, R., Raju, K.V.S.N. (2016) Fluorescence Spectroscopy of Nanofillers and their Polymer Nanocomposites, Editor(s): Sabu Thomas, Didier Rouxel, Deepalekshmi Ponnamma, Spectroscopy of Polymer Nanocomposites, pp. 158–180, William Andrew Publishing, Oxford.

[82] Straughan, B.P., Walker, S. (1976) Spectroscopy, Chapman and Hall Ltd., London, 1.

[83] Lambert, J.B., Mazzola, E.P. (2003) Nuclear Magnetic Resonance Spectroscopy: An Introduction to Principles, Applications and Experimental Methods, Pearson education Inc, New Jersey, 1.

[84] https://3.bp.blogspot.com/3M6eRY1Ufmo/UqK3PVmf09I/AAAAAAAAFsY/NY6apdtKuB0/s1600/pharmatutor-art-2076-5.png

[85] https://www.uwyo.edu/chemistry/instrumentation/nmr/_files/nmrpara.jpg

[86] Zhang, X., Dean, K., Burgar, I.M. (2010) A high-resolution solid-state NMR study on starch–clay nanocomposites and the effect of aging on clay dispersion, Polymer Journal, 42 (8), 689–695.

[87] Mojumdar, S.C., Raki, L., Mathis, N., Schimdt, K., Lang, S. (2006) Thermal, spectral and AFM studies of calcium silicate hydrate-polymer nanocomposite material, Journal of Thermal Analysis and Calorimetry, 85 (1), 119.

[88] Kaur, H., Rathore, A., Raju, S. (2014) A study on ZnO nanoparticles catalyzed ring opening polymerization of L-lactide, Journal of Polymer Research, 21 (9), 537.

Index

https://doi.org/10.1515/9783110997590-013

www.ingramcontent.com/pod-product-compliance
Lightning Source LLC
Chambersburg PA
CBHW061355210326

41598CB00035B/5989